The Didactical Challenge of Symbolic Calculators

Turning a Computational Device into a Mathematical Instrument

Springer

Mathematics Education Library

VOLUME 36

The titles published in this series are listed at the end of this volume.

Dominique Guin
Kenneth Ruthven
Luc Trouche
(Editors)

The Didactical Challenge of Symbolic Calculators

Turning a Computational Device into a Mathematical Instrument

Editors:
Dominique Guin, Université Montpellier II, France
Kenneth Ruthven, University of Cambridge, United Kingdom
Luc Trouche, Université Montpellier II, France

Library of Congress Cataloging-in-Publication Data

A C.I.P. Catalogue record for this book is available
from the Library of Congress.

ISBN 0-387-23158-7 (hardbound)
ISBN 0-387-23435-7 (E-Book)

Printed on acid-free paper.

Printed in the United States of America.

9 8 7 6 5 4 3 2 1 SPIN 11161523

springeronline.com

TABLE OF CONTENTS

ACKNOWLEDGMENTS

The French editors and non-native English-speaking authors of this book wish to express their gratitude to Kenneth Ruthven for his support in dealing with difficult language aspects.

INTRODUCTION

A significant driver of recent growth in the use of mathematics in the professions has been the support brought by new technologies. Not only has this facilitated the application of established methods of mathematical and statistical analysis but it has stimulated the development of innovative approaches. These changes have produced a marked evolution in the professional practice of mathematics, an evolution which has not yet provoked a corresponding adaptation in mathematical education, particularly at school level. In particular, although calculators -- first arithmetic and scientific, then graphic, now symbolic -- have been found well suited in many respects to the working conditions of pupils and teachers, and have even achieved a degree of official recognition, the *integration* of new technologies into the mathematical practice of schools remains *marginal*. It is this situation which has motivated the research and development work to be reported in this volume.

The appearance of ever more powerful and portable computational tools has certainly given rise to continuing research and development activity at all levels of mathematical education. Amongst pioneers, such innovation has often been seen as an opportunity to renew the teaching and learning of mathematics. Equally, however, the institutionalization of computational tools within educational practice has proceeded at a strikingly slow pace over many years. At first glance this slow pace is surprising, given the official encouragement for the uptake of new technologies by educational institutions, and the corresponding material and practical support provided by government and quasi-governmental agencies in most developed countries. Very commonly, however, the *complexity* of technology *integration* and its wider ramifications have been overlooked, underestimated or denied.

As this book will illustrate, the introduction of new computational tools calls into question established consensus concerning the mathematical knowledge to be taught in school. While some long valued components remain indispensable, others become obsolete. Technology integration also poses a significant challenge to the often sharp demarcation within schools between the domain of mathematics and that of computing or informatics. In these circumstances, systemic approaches to innovation have often displayed a high degree of caution and conservatism when it comes to technology integration. In most educational systems, for example, there has been no substantial reworking of the mathematics curriculum in more technology-aware terms. Often, indeed, the use of computational tools is barred or restricted in public examinations, creating a powerful backwash inhibiting and discouraging their uptake.

Fortunately, there are some important exceptions to this pattern. The Australian and Dutch and French teams whose work is reported in this book have all benefited from unusually favorable policies in their educational systems which have permitted -- even encouraged -- a strong integration of technology within mathematics teaching. In the French educational system, for example (Trouche, Chapter 1), the use of calculators has long been permitted (though widely contested by parents and teachers) and has increasingly been encouraged and supported by the regulations governing the baccalaureate, the public examination which marks students' completion of secondary education. Even in France, however, while tools for calculation and graphing are frequently used by students, few teachers treat them seriously as instruments for mathematical work. And these studies also demonstrate that even where educational policies are supportive the didactical complexity of integration remains a considerable barrier. Accordingly, they point to insights and mechanisms needed to assure a wider and more systematic development and diffusion of innovative approaches.

This book, then, seeks to report and synthesize a set of recent investigations with similar concerns and approaches. The projects under investigation all brought together researchers and teachers in long-term collaborations. They aimed to progressively develop teaching which exploits technology, illuminated by successive analyses of the experiments conducted. This is a realistic approach, not distancing itself from the current conditions and constraints of teaching, although it does take a view which is broader and longer than the everyday one. Consequently, the didactical designs proposed here are more an internal adaptation of current teaching than a transposition of new practices from external sources. The analyses which they offer seek to unveil the traps and obstacles produced by this incorporation of technology into teaching, so as to better understand the conditions supporting a productive integration.

Although their common focus is on the specific topic of the integration of symbolic calculators into mathematical education at upper secondary level, this topic is analyzed with the help of theoretical tools which appear to be more generally applicable without particular regard to technology type or educational level. These theoretical tools assist the researchers to take account of the delicate ecology of adaptation in a system which includes the functioning of a machine, appropriated by a user, shaped by a teaching approach, regulated by a school curriculum, framed by a wider culture. Analyzing the action and interaction of the elements of this system calls for a combination of ideas from anthropology and psychology, ergonomics and didactics, mathematics and informatics. However, the principal perspective guiding this work is that of instrumental analysis, as developed by the French research teams which provide the majority of contributions to this volume. The essentials of this perspective -- set out by Trouche (Chapter 6) -- are that, beyond the material constraints imposed by its functioning, the use of a tool depends on *schemes* which establish links between concrete *gestures* and *mathematical thinking*. In instrumental genesis, such schemes are progressively elaborated through the dual processes of the *instrumentalization of the tool* and the corresponding *instrumentation of activity*.

This perspective emphasizes the interaction of the instrumentalization of a tool with a broader instrumentation of mathematical activity, and the effect of these processes on the wider development of mathematical concepts. In particular, Lagrange (Chapter 5) signals the importance of instrumented activity in establishing a sense of mathematical objects -- as well as effecting mathematical processes -- and hence the importance of the *epistemic* -- as well as the *pragmatic* -- *function of technique*. From this point of view (as Drijvers and Gravemeijer, Chapter 7, illustrate), development of a calculator-instrumented technique may sometimes help to sensitize pupils to aspects of a mathematical concept which remain implicit in the conventional instrumentation through paper-and-pencil. In effect, the decalage between calculator and paper-and-pencil techniques may serve to stimulate an adaptation of pupils' schemes, and above all the construction of a more global scheme integrating the two techniques.

These global schemes are central to the management of conflicts between the functioning of symbolic calculators and current mathematical norms. In particular, these schemes underpin the vigilance and flexibility necessary in the face of results which can be unexpected, misleading, or even incorrect. Elbaz-Vincent (Chapter 2) reminds us that a computer algebra system does no more than treat sequences of symbols according to formal rules. Even at the syntactic level, the form and succession of the symbolic transformations carried out by the machine remain invisible to the user, which can give rise to misleading reconstructions by pupils (Lagrange, Chapter 3). Equally, the graphing algorithms of the machine produce images which need to be read more as plots than sketches of mathematical graphs (Trouche, Chapter 6). Moreover, users must take charge of the semantic aspect -- which appeals to the sense of symbols and relates them to their referents -- and the pragmatic aspect -- which takes account of wider situation and ulterior intentions.

The analyses presented in this work are sensitive to these complexities and seek to understand not only the evolution of schemes on the part of users but the variation in schemes between users. Trouche shows that, amongst calculator-using pupils, this *instrumental variation* extends from conceptions of particular operations (Chapter 6) to general orientations towards use of the machine in classroom mathematical work (Chapter 8). Equally, Kendal, Stacey and Pierce show a similar variation in usages amongst teachers (Chapter 4) combining pedagogical and mathematical aspects. Accordingly, the didactical designs proposed by the French teams seek not only to specify a series of didactical situations aimed at evoking a system of schemes on the part of the pupil, but to articulate strategies through which the collective development of these systems can be *orchestrated* and *socialized* to institutional norms (Trouche, Chapter 8; Artigue, Chapter 9).

To now move on from taking an overview of the book as a whole, to introducing each chapter in turn:

In Chapter 1, Luc Trouche offers a glimpse of the current -- primarily French -- situation as regards integration of symbolic calculators within mathematics teaching. The school as an institution faces a generation of students who wish to make use of new tools which have gained wider professional and social legitimacy; indeed they often bring such tools to the classroom. While students expect such tools to be normal *instruments* for mathematical work, a not inconsiderable proportion of teachers are reluctant to accept their integration. Such an integration supposes a more experimental conception of mathematics than that underpinning current classroom practice. The debate even extends to learned societies obliged to redefine the place of mathematics in today's world. The evolution of school programs and textbooks reflects an openness to technology integration, but evaluation of early attempts has led to a degree of institutional recognition of the difficulties of such integration.

Chapter 2, by Philippe Elbaz-Vincent, presents the general characteristics of computer algebra systems (CAS) offering the user both pre-programmed commands and a programming language which makes possible the definition of more elaborate procedures. The topic of formal differentiation is used to illustrate the types of representation provided and of manipulation made possible by such a system. A review of some classic mathematical problems serves to bring out key weaknesses of these systems, even in elementary aspects: examples are presented of results which are erroneous or difficult to interpret. In both cases, the machine response contains relevant information, but to recognize and exploit this information requires mathematical knowledge which may be at quite a high level; some examples are presented regarding the calculation of primitives in terms of standard functions. In the circumstances, what trust can the mathematician give to a proof obtained with the assistance of a CAS? More generally, the use of a

CAS calls for a critical attitude on the part of the user. Hence, to introduce such a system in teaching without thinking its use through in advance is inevitably problematic.

In Chapter 3, Jean-Baptiste Lagrange addresses in a more general way the problem of *transposition* of the professional practices of mathematicians into the teaching situation. The new tools for formal calculation make *experimental approaches* more visible and current in mathematical activity. To what extent do these new approaches provoke a broader evolution in teaching? Answering this question calls for the aims of mathematical education to be taken into account. In these changed circumstances, is it appropriate to develop in pupils something of an *algorithmic spirit*? Does using tools within an experimental approach to mathematics necessarily develop in pupils a deeper understanding of mathematical concepts? Analysis of concrete situations reveals that tools for formal calculation do not automatically support experimental work which is speedy and productive from the point of view of learning mathematics. Discussion of the conditions under which such situations are productive brings out the necessity of a strong didactic intervention in conceiving them.

Margaret Kendal, Kaye Stacey and Robyn Pierce, in Chapter 4, describe the different approaches of three Australian teachers seeking to integrate a CAS into their teaching. They underline the profound changes which this integration requires and the diversity of choices with which teachers are faced, whether these be in organization of the class, teaching of issues specifically related to use of the tool, exploiting the constraints and affordances of the tool, and managing the distribution of time involving technology and mathematics. A comparison of these teachers' approaches to integrating graphic, then symbolic, calculators makes it possible to pick out issues which are more specifically related to symbolic manipulation. It seems that styles of teaching depend strongly on the conceptions which teachers have of mathematics and on the corresponding role which they accord the calculator, and that *differences* in *styles* are *accentuated* in using computer algebra systems. Equally, the experience of these teachers encourages greater appreciation of the extent of the work necessary on the part of teachers to integrate use of CAS into their classes in a significant way. Additionally, the authors bring out the institutional problems which such innovations pose as regards curriculum and assessment.

In Chapter 5, Jean-Baptiste Lagrange takes up for consideration the new *techniques* which emerge when students appropriate tools for computer algebra. Drawing on didactical theorizations, he recalls the fundamental role which techniques play in mathematical conceptualization. Through examples taken from situations conceived by a range of research teams, he discusses the place of instrumented techniques and their interaction with more customary techniques. Instrumented techniques are not given by the tool itself, but must be thought out and put in place through situations adapted for this purpose. Consequently, important didactical choices arise in conceiving techniques and the pragmatic and epistemic functions which they will accomplish, and in determining their interaction with customary techniques. Other instrumented techniques are motivated by the need for an efficient and reasoned use of tools by students. These are clearly more difficult to make legitimate within the current institutional framework of school mathematics. This examination of the place of techniques illuminates the complexity of the teacher's role, explaining some of the diversity in approach observed in the preceding chapter.

In Chapter 6, Luc Trouche studies the processes of learning associated with the use of symbolic calculators by students. He draws attention to didactical phenomena which have been highlighted by research into the integration of graphic calculators and the software DERIVE; phenomena linked to processes of knowledge transposition or to processes of adaptation on the part of students. Then, drawing from research in cognitive ergonomics, he proposes a theoretical approach which aims to provide a better means of analyzing the distinctive process through which a technical tool is transformed into an instrument of mathematical work; an approach which makes it possible to take account of the phenomena noted earlier. A typology of constraints is set out with the help of examples relating to two different models of symbolic calculator, and these bring out the way in which particular constraints privilege certain types of action, illustrating a general method for studying an artifact. Analysis of specific examples of *instrumented techniques* related to the topic of function limits illustrates the gap between the technique taught to a class and the techniques actually practiced by students. These examples also serve to illustrate how underlying operational invariants can be inferred from techniques, and to demonstrate the fundamental role that these operational invariants play in students' conceptualizations.

In Chapter 7, Paul Drijvers and Koeno Gravemeijer draw on teaching experiments concerned with the use of symbolic calculators as tools for solving particular classes of algebraic problems, and the corresponding instrumentation of algebraic activity and correlation of new techniques with a wider algebraic discourse. The study reported in the chapter provides detailed analyses of the instrumented action schemes which students developed for solving parameterized equations and substituting expressions. It demonstrates how what might be taken simply as 'technical' difficulties often have a wider 'conceptual' aspect, so bringing out the complexity of the process of instrumental genesis. In particular, the chapter illustrates how obstacles which students encounter during the instrumentation process offer opportunities for learning, if carefully managed by the teacher, through reflection on their conceptual aspects and their relation to the corresponding paper-and-pencil technique.

In Chapter 8, taking the case of one class, Luc Trouche analyzes the diversity of processes through which a symbolic calculator becomes an instrument for mathematical work. To differentiate these processes, he establishes a typology of extreme patterns of student behavior; this typology makes it possible on the one hand to situate a particular student at a given moment in relation to the different types defined; on the other hand, to identify the development of such processes over time for a particular student. Naturally, such development depends on the situations and working arrangements put in place by the teacher. Management of the diversity of processes observed calls for the teacher to carefully plan the organization of a particular class at a given time. The theoretical approach of instrumentation is developed to describe the place of didactic intervention in the form of particular *instrumental orchestrations* defined by their objectives, architecture, and modes of operation. The didactic objective is to establish at a given moment, for each pupil and for the class as a whole, a coherent system of instruments. Different mechanisms are presented alongside one another to reinforce the social dimension of instrumented action and to manage the impact of the introduction of a new artifact on top of the *systems of instruments* which have already been established.

Chapter 9, written by Michèle Artigue, synthesizes the main contributions of the research presented in the preceding chapters. These contributions are theoretical in character: they provide a frame to problematize questions of learning and teaching in an environment of symbolic computation and to make this problematization operational. Equally, they are of an experimental character; they provide detailed information on the instrumented contexts constructed. Analysis of two pieces of didactical engineering aims to identify regularities in the choices made and the effects observed, and equally to explore the conditions for their viability. Analysis of the difficulties encountered in privileged situations is a preliminary stage in identifying the types of condition necessary for wider use. In each piece of engineering, pre-analysis of mathematical and instrumental potential precedes the detailing of the engineering intervention. Post-analysis makes it possible to measure the distance between the potentials envisaged and what actually happened in class. These descriptions and analyses show that it is possible, at least in these experimental circumstances, to construct an approach to integration, where instrumentation and mathematical knowledge are articulated with the paper-and-pencil environment, even if it does not come about by itself. These studies raise questions about the status of instrumented techniques in these experimental classes: they show the necessity of according some form of *institutional recognition* to a *coherent set of instrumented techniques* and to a theoretical discourse accompanying them.

All these insights contribute to the identification of conditions which are necessary to make viable the integration of symbolic calculators into the teaching of mathematics. They open up some lines of research which can help progress towards an effective integration which must necessarily involve institutional negotiation about mathematical needs.

Dominique GUIN is a professor at the Université Montpellier II (France), and a researcher in mathematics education. Her research in this area has dealt with the modeling of knowledge acquisition processes in mathematics.

Luc TROUCHE is an assistant professor at the Université Montpellier II (France), Director of the IREM (Institute for Research on Mathematics Teaching), and a researcher in mathematics education. His research in this area has mainly dealt with the study of conceptualisation processes in mathematics. GUIN and TROUCHE are particularly interested in the integration of computer technologies into mathematics education. Their current focus is mainly on the design of distance training, as a key means of supporting teachers in the integration of ICT and, in this context, in the conception of pedagogical resources.

Kenneth RUTHVEN is a reader at the University of Cambridge (UK) where he is Dean of Research in Education. He researches issues of curriculum, pedagogy and assessment, particularly in the light of technological change, and notably within mathematics education. His principal current focus is the teacher thinking and classroom practice associated with mainstream forms of technology use in secondary-school mathematics and science.

Chapter 1

CALCULATORS IN MATHEMATICS EDUCATION: A RAPID EVOLUTION OF TOOLS, WITH DIFFERENTIAL EFFECTS

Luc Trouche
LIRDEF, LIRMM & IREM
Université Montpellier II, France
trouche@math.univ-montp2.fr

Abstract: The appearance of more and more complex tools in mathematics classes is not a response to an institutional need of school. It is, rather, the expression within this institution of a huge social phenomenon (the increase in the number of screens and machines) arising from the utilization of computerized tools by certain branches of mathematics and science.

Alongside other computation tools, calculators have been taken into account in very different ways within the educational institution:
- students rapidly appropriate them, regarding them as of potential assistance to their mathematical work;
- teachers hesitate to integrate them in their professional practice;
- the French mathematics curriculum attempts to promote the utilization of these tools.

However, the spread of calculators raises various questions (about assessment, for example) and provokes lively discussion within professional associations.

Key words: Assessment, Computation tools, Curriculum, Mathematics evolution.

1. A SIGNIFICANT TECHNOLOGICAL EVOLUTION

1.1 Evolution of tools in Mathematics Education

For a long time, mathematics could be distinguished from other scientific disciplines by the economy and stability of the tools used in its teaching system: pencil, ruler, set square, protractor and compasses for geometry, and only pencil for computations (in western countries anyway); in Asia, other artifacts like the abacus were (and sometimes remain) widely utilized. Most probably, this apparent stability masks significant ruptures: the nature itself of the 'pencil' used for written computation may have significant effects on learning processes. Lavoie (1994) pointed out the revolution provoked by the introduction of the 'iron quill' and the pencil in Canadian schools around 1830:

> One of the reasons why the learning of writing was traditionally placed late is precisely the use of goose quills. Indeed, these tools required such dexterity to cut and use that it was normal to delay their use (...). Consequently, iron quills induced a real revolution, the learning of writing and thus of arithmetic in primary schools.

In the twentieth century, tables of numerical values and slide rules were added to these traditional tools for the scientific classes of secondary schools. Effectively, from 1925 to 1975, these tools, strongly recommended by the educational institution, were introduced by teachers and used by students for computation:

- since 1975, this situation has evolved radically. Software permitting numerical or formal computation and geometry has become accessible. The spread of small individual computing tools, calculators, has rapidly and profoundly modified students' equipment in mathematics classes. In 1975 'desk calculators' appeared, scientific and programmable calculators in 1980, graphic calculators in 1985, and symbolic calculators (provided with CAS and sometimes also with geometrical software) in 1995. When they appeared, graphic calculators cost ten times more than mathematics textbooks. Twenty years later, the cost of these two objects is similar. Tools are more and more complex, their ergonomics and performance are clearly improved, at lesser and lesser cost;
- the spread of new tools is more and more rapid. To equip all students with scientific calculators took fifteen years (from 1975 to 1990), whereas, for graphic calculators, ten years was enough (from 1990 to 2000). If this evolution carries on[1], one can anticipate most students in scientific secondary classes soon being equipped with symbolic calculators[2];

- the 'communicative' dimension of these tools is becoming more and more important. Screens have become larger and larger, devices have been designed for linking calculators together, or to a computer, or to an overhead projected calculator. Other devices have been designed for updating, via the Internet, the contents of calculator memory (flash technology), for linking a calculator with a data-logging device (Chapter 4) allowing various physical measures to be captured[3]. The affordances of these new materials do not necessarily induce utilization of these artifacts which takes a socialized form. One can notice, in fact, two convergent evolutions: computers are smaller and smaller (handheld, autonomous) and calculators are more and more integrable with communication devices.

After a long period of stability, this rapid evolution of the computing tools at students' disposal probably makes more complex the constitution of a new equilibrium in mathematics classes.

1.2 Evolution of mathematicians' tools

Software for numerical or formal computation was first introduced to meet mathematicians' needs. Mathematical practice has been deeply modified by these new artifacts. Merle (2000) identifies three main changes linked to the development of computer science in mathematics:

> - the computer has permitted, through its power of computation, the treatment of certain objects in a new light (...);
> - computerized processing raises new questions and allows certain domains to be re-examined (...);
> - [the computer induces] the expansion of discrete mathematics, applied logic and algorithmics.

More generally, the possibility of rapidly testing hypotheses facilitates the emergence of conjectures; the possibility of rapidly making a lot of computations even modifies the construction of some proofs.

One example, among many others, was suggested by Rauzy (1992). In 1770, Waring put forward the following conjecture: every integer is a sum of less than 4 squares, 9 cubes, 19 quartic powers, etc. (For example, 79 is the sum of the squares of 1, 2, 5 and 7, sum of the cubes of 1, 2, 2, 2, 3, 3, sum of the quartic powers of 1 (15 times) and of 2 (4 times)). This conjecture was proved with the assistance of a computer in the following way: Dress and Deshouillers proved that every number greater than 10^{360} was a sum of 19 quartic powers. A computer could not make all the remaining computations for smaller numbers in a reasonable time. New theoretical developments were required: it was proved that the conjecture was true if enough numbers, representable as a sum of 5 quartic powers, are in a zone close to 10^{10}. The verification required about a hundred billion operations, which could be performed by a computer in a few days... The conjecture was becoming a theorem.

Mathematical research has always included an *experimental* dimension and a conception of mathematics based on conjectures, proofs and refutations is not new (Lakatos 1976). However, due to improvement in informatics, this experimental dimension has tended to leave the private sphere of mathematicians' work and to become officially recognized (Chapter 3). Conjectures produced with the assistance of computer programs, data related to these conjectures, methods for obtaining conjectures can be explained and discussed as valid mathematical work.

Beyond this production of conjectures, there are also procedures for the investigation and definition of structures which are sufficiently general to support this production: Borwein & al (1996) show that a new scientific field is developing within the mathematical domain, whose *legitimacy, still problematic, is gaining.*

1.3 Evolution of computation tools in society

The evolution of computation tools in the practice and teaching of mathematics is also related to a profound evolution of computation tools in society. Outside the field of professional mathematicians, the professional practices of computation have, in effect, been overturned:

- in the commercial world, the appearance of computing tools induced a complete elimination of paper-and-pencil computation (it is notable that in societies where computation tools were available -- for example, the abacus -- these tools still coexist with calculators). Moreover, due to the emergence of bar codes and optical character recognition, cashiers no longer enter numbers by hand into keyboards; the only counting which remains is related to coins and banknotes so as to give change;
- the situation is quite similar for all professions involving manipulation of numbers: auditors, tax collectors and managers have at their disposal specific software for calculations linked to their professional practice;
- in the same way, *social practices of computation* have also been modified. The appearance of calculators devoted to specific tasks (conversion of national currencies into euros, cumulatively calculating the bill while shopping in supermarkets etc.), the emergence of services or software carrying standard calculations out automatically (tax amounts) may give the impression that some of the computation techniques learnt in school are now obsolete.

On the other hand, the emergence of new tools modifies not only computation techniques, but also the relationship with numbers itself. On the one hand, the taking-over of computation by a machine increases the distance of the user from numbers; on the other hand, the carrying out of marginal tasks (for instance, giving change) is connected to computation practices which involve counting concrete objects. Technological evolution dramatically changes social relationships to mathematical objects.

Chevallard (1992) brings out *the differential penetration of computer objects:*

> The spectacular penetration, still increasing, of computers in daily life and in most professional sectors hides a reality which has to be taken into account to be able to judge the stagnation in the educational domain. The degree of penetration is less a function of the way members of the institution think (modern or archaic) than of the implemented *type of use*.

Outside mathematics education, then, there is a *social legitimacy* to new computing tools, based on a clearly identified type of use, the assistance to computation.

1.4 More general evolution of tools in society

This evolution is the translation into the mathematical domain of wider evolutions. Debray (1992) points to the age of *videosphere* following the appearance of color television:

> It is the era of keyboards and screens taking the place of pencil and paper. Direct contact with things is replaced by indirect contact through the mediation of a specific machine. All that matters is what can be seen (or heard); thus the risk of confusion between an object and its representation[4], between map and territory, becomes great. In this context, the more complex and widespread use of calculators can be seen as one aspect of a wider phenomenon, marked by an increase in prostheses, the spread of screens and the miniaturization of supports.

The educational institution is thus confronted with this phenomenon both internally (owing to the introduction of calculators into classes by students) and externally (owing to the social legitimacy of these tools). This social necessity of mastering new technologies may appear then as a new legitimatization of mathematics within curricula (Schwartz 1999).

2. IN THE EDUCATIONAL INSTITUTION: THE POINT OF VIEW OF DIFFERENT PROTAGONISTS

Before examining the responses of the educational institution itself, it is probably of use to us to consider the points of view of potential actors as regards the integration of new computing tools, that is to say of their users (Baron & Bruillard 1996), students, teachers and their associations.

2.1 Students

Faure & Goarin (2001) have analyzed the results of a survey of the relationship that students have established with their calculator. This survey covered 527 students from 11th grade scientific classes. Some features emerge from these responses:

- most students (84%) own a graphic calculator (8% a calculator which is only scientific, 8% a symbolic calculator);
- the process of learning about calculators mainly takes place outside the class;
- first and foremost, the process of appropriating this tool is based on individual exploration-discovery; afterwards there is a social dimension (with friends). Generally, *the teacher is not very involved in the process of calculator appropriation*. A comparison with the results of a previous survey shows very little change over ten years (Figure 1-1).

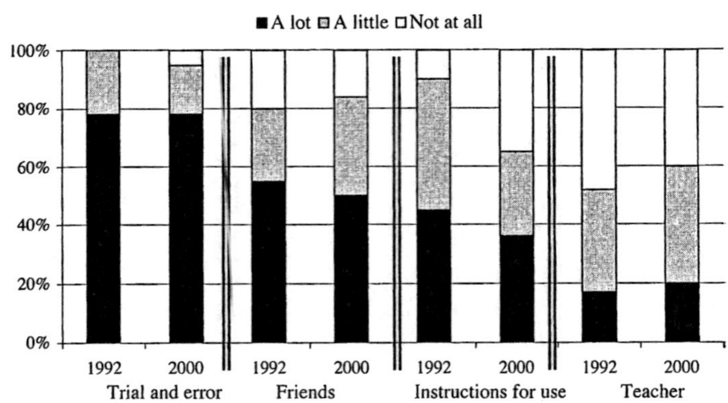

Figure 1-1. How did you learn to use your calculator?
Comparison of two surveys in 1992 and 2000 (Faure & Goarin 2001)

From further data, it appears that students hope for the teacher to be involved in the process of learning about calculators (Figure 1-2). This might be considered as an indication of institutional recognition of the tool. The actual situation, however, is the opposite of the one hoped for.

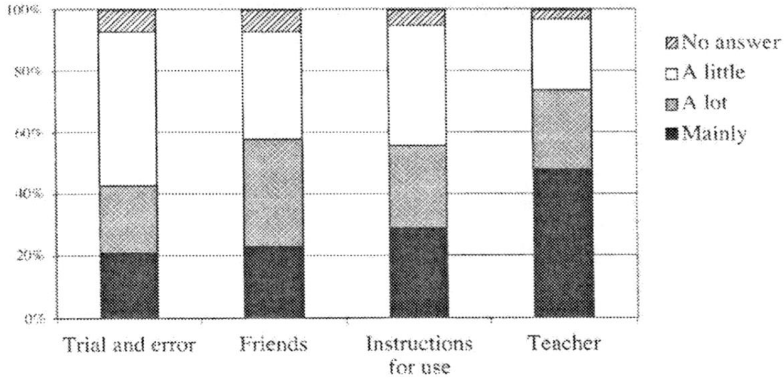

Figure 1-2. If you had to do it again, how would you prefer to learn to use your calculator? (Faure & Goarin 2001)

The authors also note (Figure 1-3) that the calculator is mainly used by students during the reinvestigation of knowledge through assessment and exercises, very moderately during more open processes of investigation and exploration, and very little when the teacher is presenting and establishing fresh knowledge.

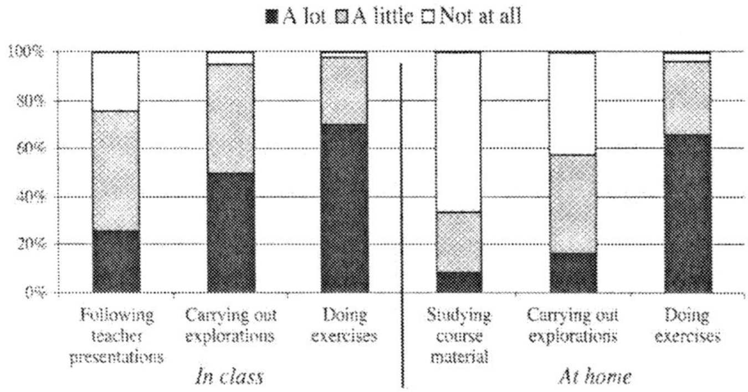

Figure 1-3. Why do you use a calculator in class or at home? (Faure & Goarin 2001)

This situation is probably linked to the weak integration of this tool into classwork.

Finally, the survey reveals that students use calculators essentially in order to graph a function, then to calculate or to study the variation of a function, and only a little to study limits or to solve equations (Figure 1-4), which reveals a quite reduced exploitation of their functionalities as tools.

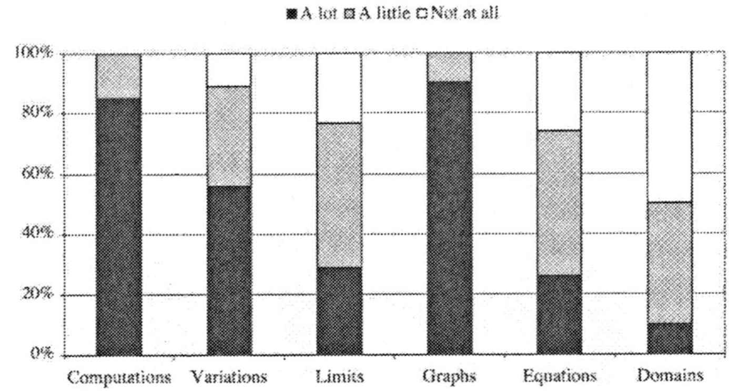

Figure 1-4. In your opinion, are calculators useful to: do computations, study variation of functions, study limits, draw graphs, solve equations, find function domains? (Faure & Goarin 2001)

In a review of new research into the integration of graphic calculators, Penglase & Arnold (1996) identify five main benefits of the use of graphic calculators which are mentioned by students, and these seem to be more important than those mentioned by Faure & Goarin:

> The ease of sketching and obtaining information from graphs; being able to check quickly the correctness of derivatives, integrals and solutions; being able to understand and interpret graphs and derivatives, integrals and solutions; the ease of computing and checking procedures regarding difficult formulae; and the increase in confidence and enthusiasm associated with the use of the tool.

However, we will see in the next chapters that the type of student work with a calculator depends on various factors:

- it depends on the teacher's role (Chapter 4). In classes where teachers do integrate the tool, most studies point to a substantial involvement of students in problem solving (Trouche 1998), with better participation in class discussions (Waits & al 1999);
- it depends on the availability of the tool (Burrill & al 2002). In this way, Chacon & Soto-Johnson (1998) point out that results are not the same when students work in computer rooms with their teacher: in this context, the authors report irritation in the face of certain tool blockages, frustration arising from the only occasional utilization of the tool. These phenomena appear rather marginally in calculator environments (i.e. when calculators are tools recognized by teachers and continuously at students' disposal);
- work varies according to the students: detailed studies, carried out in experimental contexts with specific observation devices (interviews, questionnaires, analysis of students' work), reveal the complexity of related processes in this type of environment and the possible variation in students'

behavior, according to gender (Penglase & Arnold 1996) or work method (Chapter 8).

2.2 Teachers

Available sources concerning teachers' opinions are more extensive than those concerning students (trade-union or professional sources, research results). The main feature of teachers' points of view is a reluctance to use new technologies in their teaching, at elementary school level as well as at middle school (Bruillard 1995) or secondary school level (Abboud Blanchard 1994). We can find an illustration of this situation in the survey of Faure & Goarin (ibid.), which notes that students regret not having the possibility of learning to use their calculator with their mathematics teacher.

Guin, Joab & Trouche (2003) analyzed the results of another questionnaire for teachers taking part in training courses on themes related to ICT[5]. It was not a representative selection of the community of teachers because these teachers had intentionally committed themselves to professional development (and so represent a minority). They were more disposed to question their professional practice than other teachers.

The degrees of familiarity of these teachers with calculators and computers (Figure 1-5) seem to be linked (this study is essentially related to graphical calculators, because few of the teachers know about CAS systems). But this *degree of familiarity* claimed by teachers is not necessarily linked to the frequency of utilization: Guin, Joab & Trouche highlight that, among 32 teachers indicating familiarity with their calculator, only 11 frequently use it. Moreover these similarities could mask differential patterns of response by teachers according to their calculator or computer orientation.

	Useless	Weak	Good	Very good
Computer	0	11	22	13
Calculator	0	13	23	9
CAS	19	12	8	3

Figure 1-5. What is your degree of familiarity with...
(Guin, Joab & Trouche 2003)

In this survey, the results of questions related to the type of calculator use which teachers employ with students are also interesting (Figure 1-6). It is only for quite elementary tasks (numerical work), that users are more numerous (calculators seem to be integrated there as an *ordinary tool*). For other tasks (combining *never-seldom* to one side and *sometimes-often* to the other), users and non users are equally numerous, except for quite complex tasks -- programming -- where non-users are considerably more numerous.

Moreover, it seems that very few teachers use overhead projected devices to support the integration of calculators into classes.

	Never	Seldom	Sometimes	Often
Numerical work	2	1	18	27
Algebraic work	13	8	15	10
Programming	36	5	1	2
Discovery of a notion	11	13	16	4
Illustration of a lesson result	7	18	13	6
Correction of exercises	8	16	18	2
With an overhead projected device	34	5	3	3

Figure 1-6. Type of utilization of calculators with students (Guin, Joab & Trouche 2003)

These results indicate an integration of calculators which is still weak for elementary tasks not requiring use of specific devices. Remember that these teachers were special in having chosen to take part in a training process about ICT. One may reasonably conjecture that results would not be better among the wider community of mathematics teachers. A report from the Ministry of Education about the implementation of new curricula in secondary classes (§ 3.1) also expresses these difficulties:

> (...) In spite of recurrent calls (for nearly twenty years) to take up this new tool (calculator or computer), many teachers do not feel comfortable about integrating it into their teaching[6].

The annual General Inspection report for France[7] notes, regarding utilization of new technologies, "a division between a minority of motivated teachers, frequently users, achieving high-quality results, and a majority still not involved".

This situation is not specific to France: Monaghan (1999) described a similar situation in England, estimating that "only 5% of teachers tend naturally to use new technologies in mathematics courses". However, one has to distinguish teachers' opinions about ICT in general, from those relative to their integration into classes. Bernard & al (1996) mentioned that a positive opinion about ICT is a necessary condition, but not sufficient to have a positive opinion about ICT integration.

In order to justify their reluctance, teachers often appeal to a risk of social inequality (calculators are expensive), and to pedagogical difficulties linked to the diversity of calculators used by students. However, in contexts where the same model of calculator is freely provided to all students, Bruillard (1995) pointed out that teachers remain reluctant. Consequently, we have to try to understand reasons for this mental block.

In fact, the remarks of resistant teachers express a mistrust of new tools which may have several origins:

- these tools are too crude: results, for graphic calculators, are only given approximately; this may lead to certain errors (Bernard & al 1996);
- these tools prevent some elementary learning processes; for instance, Bruillard (1995) mentions the teachers' fear that calculator use prevents the learning of the 'four operations' in middle school. One finds also a similar fear in Fromentin (1997): calculators at middle school are essential tools, but they may be dangerous;
- these tools do not fit the conception of mathematics which teachers have. One teacher explains it in this way: *"calculators deny the mathematical reflex"* (Bernard & al 1996). Reducing mathematics to an experimental practice restricts the place of formal proof.

Other reasons may explain this reluctance, not mentioned by teachers but emerging from several studies:

- the institutional discourse concerning the importance of integrating ICT has often underplayed difficulties of managing calculator environments (Guin 2001). Consequently, this discourse does not appear credible;
- indeed, the integration of complex tools into the classroom requires teachers to undertake deep questioning about their course, the exercises they have already prepared, and their professional methods. De facto, they commit themselves to a complex process of 'action research' (Raymond & Leinenbach 2000): "We conjecture that teachers who engage in action research are generally teachers who are at a critical juncture in their teaching practice and who are in a state of mind where they are open to change".

The investigation of reasons which may explain the reluctance of many teachers to integrate calculator use into their teaching establishes (in opposition) the profile of teachers favorable to this integration. Thompson (1992) and Bernard & al (1996) point out that teachers' use of ICT in mathematics classes is greater to the extent that they have an 'experimental' conception of mathematical practice (i.e. towards conjectures, refutations and proof processes). Penglase & Arnold (1996) wrote as follows:

> The data suggested that certain teaching styles are more compatible with graphic calculator use than others (…) teachers who tended to employ interactive or inquiry-oriented methodologies used the calculators during instruction more than teachers who used other teaching approaches.

The necessity of this personal conception may explain why teachers who integrate ICT into their class often remain isolated in their school (Watson 1993). It does not lead to homogeneity in teachers' practices concerning calculator use. Penglase & Arnold (1996) also noticed:

> Teachers who perceived the graphic calculator as a computational tool tended to stress content-oriented goals and viewed learning as listening. Teachers who saw it as an instructional tool had student-centered goals and discipline goals, interactive-driven teaching styles and student-centered views on learning.

We shall closely analyze this heterogeneity amongst teachers (Chapter 4). Beyond that, it is clear that most teachers express reluctance to integrating calculators into their classes. This reluctance rests on a network of reasons. To overcome them supposes, then, a set of conditions which we shall examine in the following section.

2.3 Learned associations and societies

In 1996, through the impetus given by the SMF[8], the GRIAM[9] set about combining the main French associations of mathematics practitioners. In 1998, this group defined what form calculator use might take in secondary classes:

> The great difficulty, at all teaching levels, is to arrange things so that thought lies on the student side in the student-calculator pair (…). Practically, it seems useful to explore two ways:
> 1. To carry out part of the work (exercises and assessment) without a calculator. To find, for example, the order of magnitude of solutions by mental computation (…).
> 2. To check calculator results systematically, to verify that they solve the problem and that no solution is missing (…).
> These new tools are an opportunity, and not a threat. However, the various capacities of calculators pose serious problems of equity in examinations (GRIAM 1998).

It is interesting to notice that, even if calculators are considered as *an opportunity*, the two directions given describe what could be done *before* and *after* their use, but not *with* these tools. This text is indicative of a conception which considers only, in the 'student-calculator' pair, that part of student work carried out without a calculator.

One year later, the debate on calculators was relaunched in France, in a spectacular way, by the statement of the Minister of Education, Claude Allègre: "Mathematics is being devalued, in an almost inescapable way. From now on, there are machines for computation. Likewise for drawing graphs..." (France-Soir, November 23rd 1999). The significance of this statement was such as to provoke, in 2000, a reaction from trade-union, corporate and learned associations, particularly the mathematics teacher associations, thus giving a set of opinions on the question of integration of calculation tools into mathematics:

- the Académie des Sciences[10]:

Have printing, typewriter, word-processing software or spell-checker devalued literature? More than 50 years ago, penetrating minds compared the role of informatics, just newly born, with the role of printing to valorise and develop mathematical concepts. Functions, equations, exact or approximate solutions, these notions have always depended on writing and computing tools. However, these notions have not been devalued. Today, informatics and its universal use in modeling are indissolubly linked to mathematics, which conditions their progress.

- A petition from the SNES[11] union, signed by a lot of scientists:

Developments in informatics question mathematics as much at the level of practices and tools as at the level of research domains. Mathematics education is essential in order to understand and to exploit models, to appreciate where their limits lie and what is at stake. But the place of this teaching is not determined only by the satisfaction of these needs; it is also an essential component of intellectual training in developing capacities of abstract reasoning from the moment of first contact with numbers, figures, diagrams and symbolic expressions.

- The CREM (Commission de Réflexion sur l'Enseignement des Mathématiques, Kahane 2002):

The future of the world is linked to the development of all sciences, and sciences interact today much more than yesterday. In particular, one finds concepts of mathematics and informatics in all fields of knowledge and action. Bringing mathematics and informatics into conflict is opposed not only to the history of these disciplines -- one old, the other new -- but, more seriously, it is to ignore their natural mutuality at the levels of research and use. The more powerful the means used, the more essential the mathematics. Informatics provides motivation, and a new field, for mathematics education. It leads to the revisiting of old ideas, to the introduction of new points of view and provides new food for the thought of teachers and students.

This controversy is interesting: the more recent position stressing the mutuality of mathematics and informatics reveals something of an evolution from the GRIAM text in 1998. Following this controversy, fears about the future of the discipline remain; these fears are attested to by the claim of the French Mathematical Society (SMF):

At the moment, there is a deep restlessness about the future of our discipline, while we note a weakening of mathematical training in secondary school, in content as in time allocation, and that students seem to be turning away from scientific studies.

This feeling of a discipline in peril, of a profession in danger, probably makes teachers resistant to change in general and to the integration of ICT in particular. Bottino & Furinghetti (1996) indicate that the introduction of informatics to mathematics education works only if it is perceived as a *response to a need* expressed by teachers.

As things stand, the opinions of students and teachers appear to be in overall conflict: students are quite favorable to the use of tools which they consider to be a *help for learning*, teachers are often opposed to the importation into the classroom of tools which they consider as *calling into question* their teaching.

3. INSTITUTIONAL EVOLUTIONS

Institutional evolutions may be identified directly from (formal) curriculum evolution but also from the evolution of textbooks, from the form and content of assessment and of the experiments carried out, or favored, by academic authorities. On these different points, we will attempt to set out the French situation and provide some illumination from outside.

3.1 French curriculum

The French curriculum in mathematics has evolved significantly following the 'modern mathematics' reform of the sixties and seventies which privileged the theoretical study of structures. The counter-reform of the eighties and nineties sought to favor observation, 'activities', 'problem-situations'. This counter-reform takes account of calculators. Prescriptions concerning calculation tools are displayed alongside questions relating to numerical computations and graphic representations (Box 1-1).

This box deserves several comments.

i) Since 1971, while numerical problems have played a part in most curriculum programs, those conceptual tools permitting rigorous monitoring of work with numbers have become blurred. Ideas of tolerance and error and the distinction between different number sets have disappeared. Calculator representations of number seem to exempt students from reflection on definition of objects and monitoring of results. Birebent (2001) expresses this situation as an "inability of current calculus education to resonate with numerical approximations". Since 1998, it is possible to make out something of an evolution: numerical topics have been reintroduced (only for the specialist mathematical option), in terms of the presentation of decimal, rational and real numbers. This does not settle the question of numerical approximations, but it allows the questioning of relationships between informatics and mathematics.

Box 1-1.

Evolution of curriculum (12th grade, scientific class)

Year	**Prescriptions for tools**	**Prescriptions for numerical computation**	**Prescriptions for graphic representations**
1966	Use of numerical tables of standard functions and slide rule.	Notably a specific chapter on numerical computation.	No reference to representative curve except for exponential and logarithm functions.
1971	Use of numerical tables, slide rule and computing machines.	Numerical calculations are included in the chapter. Real numbers, numerical computation, complex numbers.	The expression graphic representation appears in connection with general study of functions.
1982	Calculators will be widely used.	No chapter specific to numerical computation: it is integrated into other chapters.	Usual use of graphic representation will be promoted, because it plays a significant role in the behavior of functions.
1986	Calculators will be systematically used (a basic model is sufficient).	Numerical problems and methods play an essential role in the understanding of mathematical notions.	Graphic representations must hold a very important place in the curriculum.
1991	Calculators with statistical functions are recommended. One the other hand, graphical screens are not required.	Idem.	Idem.
1998	Graphic calculators are prescribed.	Numerical topics are introduced as an additional specialist option in mathematics.	Favoring argumentation supported by graphs.

2002	The power of investigation of computer tools and the existence of high-performance calculators, frequently at students' disposal, represent welcome progress and their impact on mathematical education is significant. This evolution has to be supported by using these tools, particularly in the phases of discovery and observation by students.	Numerical topics form a domain with which informatics strongly interacts; use of various means of computation will be balanced: by hand, with the help of a spreadsheet or a calculator.	

ii) The modification of tools is accompanied by significant modifications of the 'corresponding' mathematical field:

- the introduction of scientific calculators leads to the curriculum taking account of numerical questions: "systematic use of calculators reinforces the study of numerical questions, as much to make calculations, as to check results or to support research work" (curriculum program, 1986);
- the introduction of graphic calculators leads to graphical representations being taken more and more into account: "A geometrical vision of problems will be developed in calculus, because geometry supports intuition with its language and its representation procedures" (curriculum program, 1994). Consequently, we should notice that the *graphic frame* is privileged in the calculus part of curriculum, as for example in the following comment[12]: "Deeper work is suggested on the limit of sequences, easier to tackle than the function limit at a point: the objective is ambitious, so it is advisable to remain reasonable in implementing it and to favor arguments supported by graphs" (DESCO 2001). One may conjecture that the type of tool favored has an influence on the frame of work: with scientific calculators, the numerical frame is favored, whereas it is the graphical one with a graphic calculator;
- a readjustment between numerical and algebraic may be pointed out in the curriculum of the 12th grade scientific class (2002): "An approach as much numerical (with calculation tools: calculator or computer) as graphical or algebraic will be adopted".

iii) Prescriptions relating to calculators are more and more precise: they speak of *use*, then *wide use*, then *systematic use* (Box 1-1). One may also distinguish three stages in the relationship between development of tools and curriculum prescriptions:

- first period (1971-1982): the choice is made to introduce tools having a social legitimacy (calculating machines) into school, but this decision has no practical consequence (educational establishments do not tool up, teachers do not respect prescriptions);
- second period (1982-2000): the choice is made to *integrate tools so as to be at students' disposal, following, but lagging behind,* the actual availability of equipment (in this way, graphic calculators were prescribed in 2000, whereas for five years all students had already had one (§ 1.1));
- third period (since 2001), the choice is made to *anticipate the equipment* available to *students and establishments* and to prescribe the use of tools which significantly modify mathematical work (dynamic geometry software, spreadsheets, CAS). It seems that this behavior aims to overcome teachers' reluctance (§ 2.2). In 2001, for the first time in the curriculum of 11th grade scientific classes, specific devices integrating computer tools are mentioned: "The curriculum does not fix any distribution between different modes which should all be present: student activities with computer or programmable graphic calculators, work with the whole class (or in small groups) using a computer with an overhead projected device" (DESCO 2001).

iv) Finally, an evolution can be noted in the type of integration sought by curriculum designers: in 1971, tools were considered as an assistance for computation; since 1982, they have been used to favor behaviors of observation, conjecture and checking; since 2001, tools have become essential elements of mathematical work at all levels (conjecture and checking, but also in *forming a spirit of rigor*): "the student must learn to place and integrate use of computer tools into a purely mathematical process" (curriculum program of 11th grade scientific class, 2001).

This evolution illustrates the difficulty of the institutional choices necessary in order to integrate new tools into the class: to lag or lead the state of students' individual equipment (with financial repercussions: complex tools are expensive), following or anticipating changes in teachers' practices, stressing more the 'direct' use of tool or its mathematical control, favoring one application (graphic, numerical or symbolic) or the articulation between different applications, introducing programming or using calculation programs already implemented, etc.

3.2 Curriculum in the USA

These problems of integrating calculators into mathematics education are posed in all educational systems, with a great variety of responses (Drijvers 1999). In the USA, Waits et al (1999) remind us that the NCTM (National

Council of Teachers of Mathematics) Curriculum and Evaluation Standards (published in 1989) emphasize how accessible graphic calculators can be to students because they are handheld computers which may be carried around in a bag or pocket.

It is interesting to notice that, in 2000, the NCTM confirmed this technological choice:

> Calculators and computers are reshaping the mathematical landscape, and school mathematics should reflect those changes. Students can learn more mathematics more deeply with the appropriate and responsible use of technology. They can make and test conjectures. They can work at higher levels of generalization or abstraction.

... But the NCTM emphasizes, nevertheless, the necessary *control by teachers* of the integration process:

> Technology cannot replace the mathematics teacher, nor can it be used as a replacement for basic understandings and intuitions. The teacher must make prudent decisions about when and how to use technology and should ensure that the technology is enhancing students' mathematical thinking.

A similar evolution may be noticed in France:

- when graphic calculators appeared, the illusion was conveyed that students would easily appropriate these 'handheld computers' as an instrument for their mathematical work;
- at a later point, the responsibility of teachers for the integration of tools into classes is emerging. The report of IGEN (2000), previously quoted, notes: "Far from withdrawing in favor of face-to-face contact between student and machine, teachers are required to play a significant role, certainly a modified one, but still a determining one. They are mediators of the access to knowledge and training".

3.3 Evolution of textbooks

The study of textbooks is interesting because it allows the gap to be estimated between official curriculum prescriptions and their application by teams of teachers writing textbooks. For twofold reasons, textbooks are situated in an intermediate position between what was formerly taught and what is now to be taught:

- old programs have put their mark on writers who have often already written textbooks for them;
- publishers prefer that teachers not be confronted with anything too daring; even in the case of significant change in the curriculum program, they want to achieve elements of continuity.

A study of textbooks in different countries reveals a broad spectrum concerning the use of calculators. It is interesting to pick out extremes:

- in Japan, calculator use is extremely limited. Accordingly, Eizo (1998), analyzing results from the Third International Mathematics (and Science) Study (TIMSS), notes:

> When Japanese mathematics textbook is compared with the international standard, it is characterized as follow: there is a little consideration of individual ability, topics already learned are rarely repeated, and there are a few situations in which calculators are used. And it is pointed out that algebra and proof in geometry are emphasized in the second grade of lower secondary school.

- In the USA, a textbook (Demana & al 1997) integrating graphic calculators provides a reference point. In contrast to Japanese mathematics textbooks, this one calls widely for student activities, observation and conjectures. On the other hand, this textbook requires very little process of theoretical validation (we should note that the publication of this book closely followed the appearance of graphic calculators, and so is situated before the evolutions indicated in § 3.1);
- in France, new textbooks (2002, for the 12th grade class) give a much more significant place to graphic calculators and geometry software (screen images inserted in the text, hints made for use, specific activities suggested). From now on, textbooks will be accompanied by CD-ROMs providing animations. In 2001, for the first time, textbooks refer to symbolic calculators (Box 1-2).

3.4 The issue of assessment

The issue of assessment is complex; it can be tackled from different points of view:

- how the educational institution chooses to assess the use of tools it prescribes;
- which choices are made by different educational systems;
- what types of exercise are considered basic in order to assess the mastery of a given tool.

Box 1-2.

Extract of a textbook

(11th grade, Bontemps 2001, p.65)

Calculators allow checking of the computation of f ' (a) for a particular value of a in different ways:

a) *Numerically*, computing the rate of change $\frac{f(a+h)-f(a)}{h}$ for 'small' h. This gives an approximate value for f ' (a). Using the command *nDeriv* directly gives an approximate value of f ' (a).

b) *Graphically*, considering the graph of f at the neighborhood of a: the slope of the tangent at *a* is equal to f ' (a).

c) *Formally*, using a CAS (available on some calculators). Such software gives the exact value of f ' (a).

Example: Let the function f be defined on]0 ; +∞[by f(x) = $\frac{1}{1+x}$. Calculate f ' (2) 'by hand', then check the result with a calculator.

1) *Computing 'by hand'* $\frac{f(2+h)-f(2)}{h} = \frac{\frac{1}{3+h}-\frac{1}{3}}{h} = \frac{\frac{3-3-h}{3(3+h)}}{h} = \frac{-h}{3h(3+h)} = \frac{-1}{3(3+h)}$.

Then f ' (2) = $-\frac{1}{9}$.

2) *With a calculator*

a) *Numerically*

One calculates the rate of change between 2 and 2 + h, for a 'small' value of h (below, h = 0.01). An approximate value of f ' (2) is – 0.11.

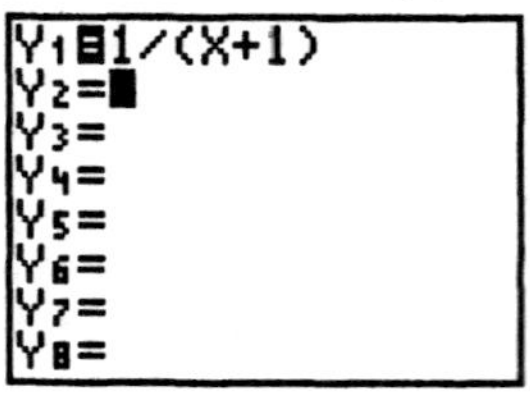

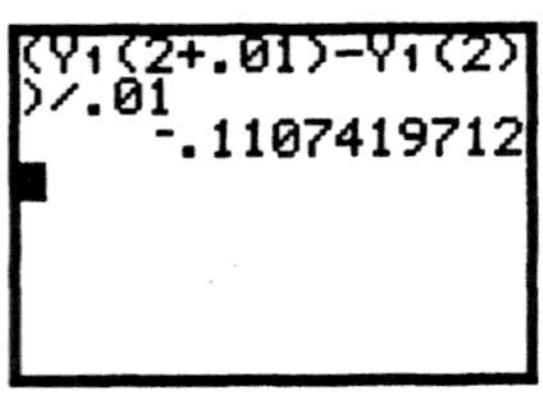

On a TI-82, the command *nDeriv* directly gives an approximate value of f ' (a): one types the function, the variable and the value of *a*. One may also indicate, or not, a particular value for h.

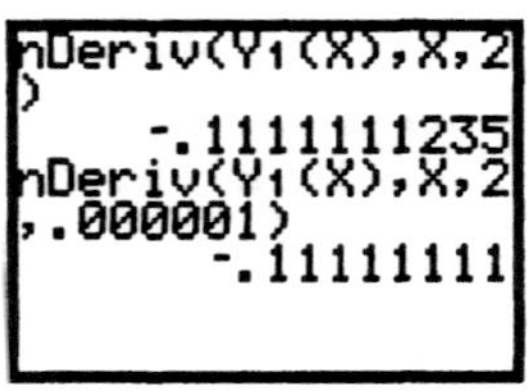

b) *Graphically*

One considers the (cartesian) graph of f in the neighborhood of 2 (one may try successive zooms). The slope may be estimated or calculated by referring back and adjusting the window. See below an estimate of the slope of the tangent equal to $\frac{y\,max - y\,min}{x\,max - x\,min}$:

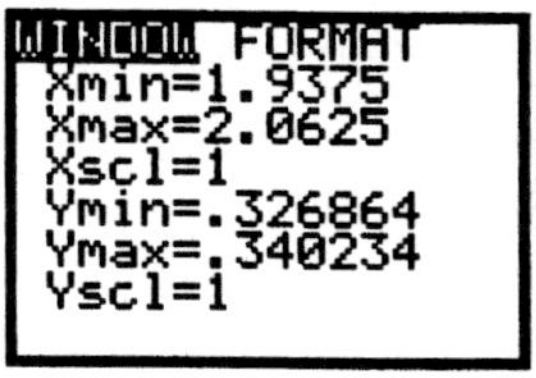

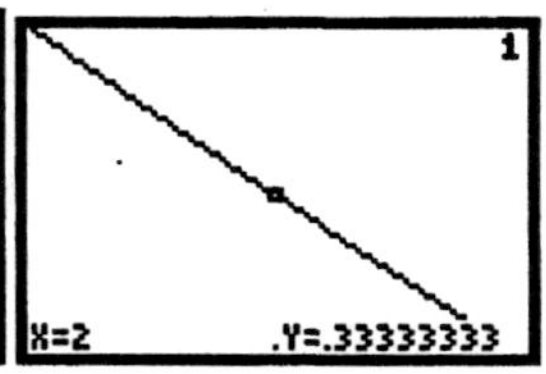

c) *Formally*
Only a symbolic calculator (see below, a TI-92) gives the *exact* value of f ' (2).

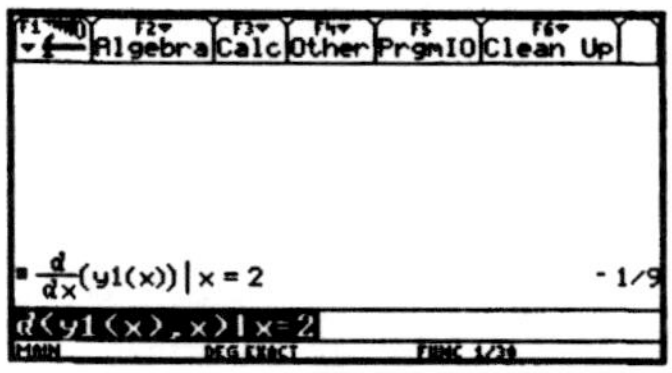

3.4.1 About institutional assessment

Institutional assessment is interesting from two points of view:
- it indicates what an institution considers it necessary for students to know, allowing the 'hard core' of the curriculum and related official prescriptions to be determined;
- it allows eventual uncertainties (corresponding, for example, to what an institution requires to be taught, but not to be assessed) to be identified.

In France, we focus on the level of the baccalauréat, a secondary school examination conferring a university entrance qualification. All types of calculators (scientific, graphic or symbolic) are allowed in this examination. The corresponding ministerial circular stipulates, in 1995:

> Mastery of calculator use represents a significant objective in the training of students because it constitutes an efficient tool within the context of their studies and their professional, economic and social life. Therefore, the utilization of calculators is provided for in numerous educational programs and they must be widely used in examinations. The equipment allowed comprises all pocket calculators including programming, alphanumeric or graphic calculators, provided that their running is autonomous and that no printer is used.

This circular replaces an older one (1986) which restricted only the surface area of calculators:

> All pocket calculators, programming, alphanumeric calculators included, are allowed provided that their running is autonomous and that no printer is used. In order to limit machines to a reasonable format, their base surface must not exceed 21 cm x 15 cm.

With different wording, successive circulars have thus authorized all types of calculators for the baccalauréat in France. This stability should not be allowed to mask some institutional difficulties:

- whereas, in this examination, all types of calculators are allowed, most French universities prohibit them in examinations;
- the institution still does not know how to integrate calculator use during the examination. For example, whereas in 1998 (§ 3.1) programs prescribed for the first time the utilization of graphic calculators in 12th grade, a ministerial circular, issued just after the start of the new school year, explained that the baccalauréat would not include any question requiring the use of these tools.

This revision attests to the hesitancy of the institution confronted with criticism from the media, and with the significant resistance of some teachers[13]. We should note that the examination in June 2000 did not include any question explicitly requiring graphic calculator use.

The baccalauréat is the keystone of secondary school education in France; consequently these controversies were sharp ones within the institution. In this context, one can thus understand that changes are more difficult to institute.

3.4.1.1 Comparison between different institutional choices

Drijvers (1999) carried out a comparative study in several countries (Germany, England, Denmark, France, the Netherlands). The author drew up a typology of different choices of assessment:

- *technology partially prohibited* (Denmark was preparing such a strategy for its final national examination). This examination is divided into two tests: for the first one, calculators are prohibited; for the second one, calculators (and all type of documents, books, exercise books) are allowed;
- *technology allowed, but benefits avoided* (for example, in England, during the final examination, graphic calculators are allowed). In this context, the curve is given in the examination paper, and then graphic calculators do not provide any direct benefit;
- *technology recommended and useful, but no added marks* (Drijvers places the baccalauréat in this category). Symbolic calculators are permitted, and they allow, for example, required limits of functions to be determined. However, if candidates do not justify the given result by a theorem or adequate reasoning, their response will not be given any marks;
- *technology obligatory and rewarded* (for example, in the Netherlands, a final examination in experimental classes). Students who determine an approximate value of the maximum of a function, thanks to their graphic calculator, obtain part of the marks attached to this question, because they could translate the question into the calculator language and knew how to select the relevant commands.

Thus, in all the countries mentioned (outside the context of experimental classes), the objective of assessment seems to be to bypass the calculator's existence.

3.4.1.2 What assessment can be imagined for calculators?

Aldon (1997) distinguishes four types of possible exercises for assessment in a symbolic calculator environment:

- exercises aiming to test the ability to interpret an answer or a non-answer from a machine;
- exercises aiming to test the ability to put forward a conjecture, to explore a domain;
- exercises aiming to evaluate the ability to decide on the usefulness of the machine in a given situation;
- exercises aiming to test a student's ability to use different frames in order to conjecture or validate an answer.

We can see, in institutional thinking, the same interweaving between calculator use and the implementation of 'experimental' behavior: in 1996, a committee specific to the baccalauréat was set up in France by the Ministry of Education. Several periods can be identified:

- in 1996-97, this committee focused on writing new examination questions and on the implementation of specific devices allowing work to be evaluated with and without a calculator. A twofold test was experimented with: a first part with the calculator, a second part without the calculator (the committee also had two other ideas: all calculators prohibited, or the definition of an approved calculator to be used in the baccalauréat);
- during 1997-98, following a request from associations of mathematicians, the committee extended its work to find more 'open' questions, allowing real research during examinations (Guin & Trouche 2002, p. 109);
- since 1998, whereas the ministry seems to have given up the previous experiment, considering it was not very convincing, the committee has focused its work on developing examination questions allowing students to take initiatives.

These evolutions seem significant from two points of view:

- the integration of ICT into mathematical work necessarily requires the implementation of a more experimental behavior (§ 1.2 and Chapter 3);
- it is difficult to integrate a given tool into assessment before having integrated it into the corresponding teaching.

3.5 Experiments

In spite of the reluctance of most teachers, numerous ways and local experiments have been developed for integrating calculators into mathematics classes.

3.5.1 Experiments piloted by the educational institution

In France, since 1991, the Ministry of Education has piloted experiments aiming to integrate CAS (Hirlimann 1998), then symbolic calculators (after 1996): all students of experimental classes were provided with the same calculator (TI-92). These experiments were carried out by university research teams (DIDIREM at Paris, ERES at Montpellier, IUFM at Rennes). Results (Guin & Delgoulet 1997, Trouche 1997, Artigue & al 1998) point out potentialities and constraints of this type of environment. The fact that the educational institution has piloted these experiments reveals the importance attached to this environment. But, until now, the results of these experiments have been only marginally exploited, particularly in the IREMs (Box 1-3), teacher training or other academic sites. New experiments are going on, using new possibilities of connection with new equipment:

Box 1-3.

The network of IREMs

Instituts de Recherche sur l'Enseignement des Mathématiques

(Research Institutes on the Teaching of Mathematics)

IREM were created within universities in France, at the beginning of the seventies. At the time, their official assignment was to assist teachers with the 'modern mathematics' reforms. From year to year, this assignment has evolved. At present, the IREMs constitute active educational research centers, based on teamwork in which academics join with primary and secondary teachers, combining experiments in classes with more theoretical researches.

The IREMs' contribution was particularly significant in promoting reflection on the integration of calculating tools into mathematics education:

- many experiments mentioned in this book (Chapters 5, 6, 8 and 9) were carried out within these institutes;
- numerous resources (documents for classes, in-service and pre-service teacher training) have been developed in the IREMs.

Resources produced by these institutes can be found at: (http://www.univ-irem.fr).

- in the academy of Bordeaux where a device networking symbolic calculators, linked to a computer, connected to the Internet and to a video-projector, has been tested in two classes;

- in the academy of Montpellier (Conclusion Chapter) where a distance training and assistance device was developed for mathematics teachers. This specific device has been ongoing since September 2000 with several themes including integration of symbolic calculators (Guin, Joab & Trouche 2003).In other countries, other experiments toward the integration of symbolic calculators are in process, particularly in Austria (Schneider 1999), in Australia (Kendal & Stacey 2001) and in the USA (Waits & al 1999).

In Luxembourg, since November 2003, all students of scientific classes have been provided by the academic authorities with a symbolic calculator. For the first time, a whole country has decided to modify mathematical learning environments. The Luxembourg Ministry was conscious of the difficulties of this enterprise and decided to carry out this experiment in the light of the previous experiments in France and in the Netherlands. An European project is in progress including reflection about efficient assistance to this national experiment.

3.5.2 Experiments encouraged by manufacturers

Calculator manufacturers (particularly Casio and Texas Instruments) maintain a continuous presence to mathematics teachers: regular delivery of periodicals to all secondary teachers, organization of meeting days for presenting new equipment, proposing activities on dedicated websites, presence at all pedagogical meetings, financial support for specific conferences, recruiting in-service mathematics teachers to present new equipment.

It is interesting to notice the evolution of the discourse employed by manufacturers:

- when graphic calculators first appeared, the approach was enthusiastic and naïve. In a discourse aimed at students, the idea was that the tool, which is easy to pick up, allows computation to be facilitated and learning difficulties to be solved; on the other hand, a discourse aimed at teachers emphasized the potentialities of the tool to capture students' attention, to represent curves, to make the class lively and efficient. Calculators were presented as *natural instruments* of mathematical work;
- when symbolic calculators appeared, due to the increasing complexity of tools, and perhaps seeing teachers' reluctance to take up graphic calculators (§ 2.2), manufacturers adopted a discourse which emphasizes more the *support necessary* for integration of tools (suggestion of activities and training meetings). Integration of calculators is presented as a strategy in renewing mathematics teaching, requiring mobilization of teachers. Texas Instruments (TI) provides the most illuminating example with its T^3 organization (Teachers Teaching with Technology) which has developed a

program of research and teacher training (for TI technology) in different countries[14]. However, the emphasis remains on the potentialities as a tool, even if difficulties or constraints are also mentioned[15].

This concern to come closer to teachers may even lead to experiments being proposed involving the integration of calculators into classes. In this way, in 1998, TI equipped about thirty French classes with symbolic calculators (TI-89). Teachers of these classes were TI trainers; thus, they knew the TI-89 well and were entirely favorable towards their use in class. However, one year later, the results were very mixed:

- for some teachers, the experiment was interesting: Benzérara & Guillemot comment (2001): "It's a very enriching experiment in which we have learnt many things";
- most teachers were amazed at the extent of the work necessary, and the numerous practical difficulties which had to be overcome to integrate calculators in their class (Noguès & Trouche 2000). Even the presence of an expert teacher (in mathematics and in new technologies), favorable to these new technologies, and with an experimental conception of mathematics (§ 1.2), does not guarantee the success of such an experiment. Explaining the potentialities of new tools and actually integrating them into a class are not similar tasks.

4. CONCLUSION

Various experiments show that the integration of tools into schooling requires specific strategies and deep reflection. This awareness emerges also in the writings of authorities reflecting on the long-term evolution of the educational system. In France, the trends of new curricula are defined at a national level. Artigue, responsible for the computation theme, brings out the crucial issue of instruments:

> This question of instruments in mathematical activity is without doubt a crucial question today and it arises from the start of elementary school up to the university. The domain of calculation is particularly perceptible because a narrow vision of it may lead people to think that it is no longer necessary to learn, because instruments can today take charge of a part of technical work which was previously devolved to us. We have to prove that, even if new balances must emerge and mathematical needs are evolving, an instrumented intelligent, efficient and controlled mathematical practice requires significant mathematical knowledge (Kahane 2002).

This evolution is not peculiar to France. A review of recent research at international level (Lagrange & al 2003) points out a similar evolution in most countries where questions of ICT integration have been posed. After a period of enthusiasm, then another of hesitation, a time seems to come of awareness of the difficulties of integrating new computing tools, in educational institutions and at the research level. These difficulties are at the heart of the problems that confront the institution (problems of social equality, of the number of students in classes, of teacher training, of the development of mathematics, of curriculum change, of the status of experimental mathematics, of individual assistance, of student research work). On all these points, the integration of calculators opens new possibilities and raises new problems.

NOTES

1. This fact depends on the one hand, on the permission, or not, of these tools in examinations; on the other hand, it depends on the reactions teachers will have about tools which deeply question their professional practice.
2. Each year, according to the manufacturers, about 600,000 graphic calculators are sold in France, and each year this number grows by 5% (compared with the number of students in 10th grade classes: 500,000). 20,000 symbolic calculators were sold in 1997, 40,000 in 1998, 60,000 in 1999 (compared with the number of students in scientific classes following a mathematical option: 60,000).
3. For example, the CBL (Calculator-Based Laboratory) of Texas Instruments, allowing (according to TI) real world data to be collected, thus the world of math and science to be explored by students.
4. One can understand the success of graphic calculators: when they first appeared, these machines were named programmable graphic calculators. Very quickly they were named only as graphic calculators, because the main functionality for users was to represent curves.
5. ICT: Information and Communication Technologies.
6. June 2000, *Bilan de la mise en œuvre anticipée durant l'année scolaire 1999-2000 du programme de mathématiques de la classe de seconde*, Ministère de l'Education Nationale.
7. 2000, official survey, Ministère de l'Education Nationale.
8. French Mathematical Society.
9. Literally, the Group for Reflection between Associations in Mathematics.
10. Academy of Sciences.
11. Syndicat National des Enseignements du Second degré (National Union of Secondary School Teachers).
12. 11th grade scientific class, 2001.
13. Moreover in 1999, a petition by electronic mail was set up by mathematics teachers, in order to ask for the prohibition of calculators in the baccalauréat.
14. A description of the T^3 organization in the USA can be found at (www.ti.com/calc/docs/t3info.htm), for France at (www.ti.com/calc/france).
15. In a Texas Instruments publication introducing CAS, (Fortin 1998) mentioned: "The use of any tool, even a purely numerical one, may cause errors linked to the method of manipulating data. Only a minimal knowledge of some particularities of the working of calculators is needed to correctly interpret them". This idea will be developed in Chapter 2.

REFERENCES

Abboud-Blanchard M. (1994) *L'intégration de l'outil informatique à l'enseignement secondaire des mathématiques: symptômes d'un malaise*, Doctoral Thesis. Paris: Université Paris VII.

Aldon G. (1997) A propos de l'évaluation, in IREM de Rennes (Ed.), *Calcul formel et apprentissage des mathématiques, Actes de l'université d'été 1996* (pp. 221-226). Rennes: IREM, Université Rennes I.

Artigue M., Defouad B., Duperier M. Juge G. & Lagrange J.B. (1998) Intégration de calculatrices complexes à l'enseignement des mathématiques, *Cahier DIDIREM spécial* n° 3. Paris: IREM, Université Paris VII.

Baron G.L. & Bruillard E. (1996) *L'informatique et ses usagers dans l'éducation*. Paris: PUF.

Benzérara M. & Guillemot A. (2001) Deux ans en Terminale S avec la TI-89, in J.B. Lagrange & D. Lenne (Eds.), *Calcul formel et apprentissage des mathématiques, Actes des journées d'étude Environnements informatiques de calcul symbolique et apprentissage des mathématiques* (pp. 171-175). Paris: INRP.

Bernard R., Faure C., Noguès M. & Trouche L. (1996) L'intégration des outils de calcul dans la formation initiale des maîtres, *Rapport de recherche IUFM-MAFPEN*. Montpellier: IREM, Université Montpellier II.

Birebent A. (2001) *Articulation entre la calculatrice et l'approximation décimale dans les calculs numériques de l'enseignement secondaire français*, Doctoral Thesis. Grenoble: Université Joseph Fourier.

Bontemps G. (Ed.) (2001) *Manuel de la classe de première scientifique*, Collection Fractale. Paris: Bordas.

Borwein J., Borwein P., Girgensohn R. & Parnes S. (1996) Making Sense of Experimental Mathematics, *The Mathematical Intelligencer* 18(4), 12-18.

Bottino R. & Furinghetti F. (1996) The emerging of teachers' conceptions of new subjects inserted into mathematics programs: the case of informatics, *Educational Studies in Mathematics* 30, 109-134.

Bruillard E. (1995) Usage des calculatrices à l'école élémentaire et au début du collège, *Rapport de recherche*. Créteil: IUFM de Créteil.

Burrill G.J., Allison J., Breaux G., Kastberg S., Leatham K. & Sanchez W. (2002) *Handheld Graphing Technology in Secondary Mathematics: Research Findings and Implications for Classroom Practice*. Texas Instruments.

Chacon P.R. & Soto-Johnson H. (1998) The Effect of CAI in College Algebra incorporating both Drill and Exploration, *The International Journal of Computer Algebra in Mathematics Education*, 201-216.

Chevallard Y. (1992) Intégration et viabilité des objets informatiques, in B. Cornu (Ed.), *L'ordinateur pour enseigner les mathématiques* (pp. 183-203). Paris: PUF.

Debray R. (1992) *Vie et mort de l'image, une histoire du regard en Occident*. Paris: Gallimard.

Demana F., Waits B.K. & Clemens S.R. (1997) *Precalculus, a Graphing Approach*. USA: Addison-Wesley Publishing Company.

DESCO (Ed.) (2001) *Accompagnement des programmes - Mathématiques - Classes de première des séries générales*, CNDP.

Drijvers P. (1999) Evaluation et nouvelles technologies: différentes stratégies dans différents pays, in D. Guin (Ed.), *Calculatrices symboliques et géométriques dans l'enseignement des mathématiques, Actes du colloque francophone européen* (pp. 127-157). Montpellier: IREM, Université Montpellier II.

Eizo N. (1998) Curriculum Studies Based on International Comparison of Mathematics Education, *Towards the reform of curriculum in mathematics*. Tokyo: Society of Mathematical Education.

Faure C. & Goarin M. (2001) *Rapport d'enquête sur l'intégration des technologies nouvelles dans l'enseignement des mathématiques en lycée*. Montpellier: IREM, Université Montpellier II.

Fortin P. (1998) Introduction au calcul formel, T³ Europe: document Texas Instruments.

Fromentin J. (1997) La calculatrice au collège, in P. Legrand (Ed.), *Profession enseignant, les maths en collège et en lycée* (pp. 150-166). Paris: Hachette.

GRIAM (1998) Lycées, quels programmes pour quels objectifs? *Bulletin de l'APMEP* 414, 103-117.

Guin D. (2001) Intégration des outils de calcul symbolique dans l'enseignement des mathématiques: comment concevoir une formation mieux adaptée? *Le métier d'enseignant de mathématiques au tournant du XXIème siècle, Actes de l'Université d'Eté* (pp. 77-93). Paris: APMEP.

Guin D. & Delgoulet J. (1997) *Etude des modes d'appropriation de calculatrices graphiques et symboliques dans une classe de seconde*. Montpellier: IREM, Université Montpellier II.

Guin D., Joab M. & Trouche L. (2003) *Suivi de formation à distance pour les enseignants de mathématiques, bilan de la phase expérimentale (CD-ROM)*. Montpellier: IREM, Université Montpellier II.

Guin D. & Trouche L. (2002) (Eds.) *Calculatrices symboliques: transformer un outil en un instrument du travail mathématique: un problème didactique*. Grenoble: La Pensée Sauvage éditions.

Hirlimann A. (Ed.) (1998) *Faire des mathématiques avec un système de calcul formel*, CNDP: Champagne-Ardenne.

IGEN (2000) Rapport de l'Inspection Générale de l'Education Nationale, Ministère de l'Education Nationale.

Kahane J.P. (2002) *L'enseignement des sciences mathématiques*, Commission de réflexion sur l'enseignement des mathématiques, Centre National de Documentation Pédagogique. Odile Jacob.

Kendal M. & Stacey K. (2001) The impact of teacher privileging on learning differentiation with technology, *International Journal of Computers for Mathematical Learning* 6(2), 1-23.

Lagrange J.B., Artigue M., Laborde C. & Trouche L. (2003) Technology and Mathematics Education: a Multidimensional Study of the Evolution of Research and Innovation, in A.J. Bishop, M.A. Clements, C. Keitel, J. Kilpatrick & F.K.S. Leung (Eds.), *Second International Handbook of Mathematics Education* (Vol. 1, pp. 239-271). Dordrecht: Kluwer Academic Publishers.

Lakatos I. (1976) *Proofs and Refutations, the Logic of Mathematical Discovery*. Cambridge: University Press.

Lavoie P. (1994) *Contribution à une histoire des mathématiques scolaires au Québec: l'arithmétique dans les écoles primaires (1800-1920)*, Doctoral Thesis. Laval: Faculté des sciences de l'éducation, Université de Laval, Québec.

Merle M. (2000) Mathématiques et informatique, informatique et enseignement des mathématiques, *Bulletin de l'APMEP* 429, 449-454.

Monaghan J. (1999) Les enseignants et la technologie, in D. Guin (Ed.), *Calculatrices symboliques et géométriques dans l'enseignement des mathématiques, Actes du colloque francophone européen* (pp. 159-163). Montpellier: IREM, Université Montpellier II.

Noguès M., Trouche L. (2000) Quelle prise en compte des contraintes didactiques dans des environnements technologiques complexes? In M. Bailleul (Ed.), *Les instruments dans la pratique et l'enseignement des mathématiques, Actes de l'école d'été de didactique des mathématiques* (pp. 214-234). Caen: IUFM.

Penglase M. & Arnold S. (1996) The graphics calculator in mathematics education: a critical review of recent research, *Mathematics Education Research Journal* 8(1), 58-90.

Rauzy G. (1992) L'informatique conduit-elle à des mathématiques nouvelles? In B. Cornu (Ed.), *L'ordinateur pour enseigner les mathématiques* (pp. 97-117). Paris: PUF.

Raymond A.M. & Leinenbach M. (2000) Building a Versatile Understanding of Algebraic Variables with a Graphic Calculator, *Educational Studies in Mathematics* 41(3), 283-307.

Schneider E. (1999) La TI-92 dans l'enseignement des mathématiques, des enseignant(e)s découvrent la didactique des mathématiques, in D. Guin (Ed.), *Calculatrices symboliques et géométriques dans l'enseignement des mathématiques, Actes du colloque francophone européen* (pp. 49-60). Montpellier: IREM, Université Montpellier II.

Schwartz J.L. (1999) Can technology help us make the mathematics curriculum intellectually stimulating and socially responsible? *International Journal of Computers for Mathematical Learning* 4, 99-119.

Thompson A.G. (1992) Teacher's beliefs and conceptions: a synthesis of the research, in D.A. Grouws (Ed.), *Handbook of research on mathematics teaching and learning* (pp. 127-146). New York: Macmillan.

Trouche L. (1997) *A propos de l'enseignement des limites de fonctions dans un environnement calculatrice, étude des rapports entre processus de conceptualisation et processus d'instrumentation*, Doctoral Thesis. Montpellier: Université Montpellier II.

Trouche L. (1998) *Expérimenter et prouver: faire des mathématiques au lycée avec des calculatrices symboliques*. Montpellier: IREM, Université Montpellier II.

Waits B, Longhart K. & Longhart F. (1999) Le rôle des calculatrices symboliques dans la réforme de l'enseignement des mathématiques, in D. Guin (Ed.), *Calculatrices symboliques et géométriques dans l'enseignement des mathématiques, Actes du colloque francophone européen* (pp. 39-47). Montpellier: IREM, Université Montpellier II.

Watson D.M. (Ed.) (1993) *The Impact Summary*, DFE and King's College London.

Chapter 2

A CAS AS AN ASSISTANT TO REASONED INSTRUMENTATION

Philippe Elbaz-Vincent
UMR CNRS & Département de Mathématiques
Université Montpellier II, France
pev@math.univ-montp2.fr

Abstract: We propose to illustrate an approach to *reasoned instrumentation* with a Computer Algebra System (CAS). We give an overview of how a CAS works (in our case MAPLE), and also some explanations of the mathematical theories involved in the algorithms described. We point out the failures of some methods, and how we can anticipate and prevent them. We conclude by giving some personal insights on how we can use a CAS in an educational framework.

Key words: Computation of limits, Computer Algebra System, Reasoned instrumentation, Symbolic integration.

1. PRELIMINARIES

1.1 Foreword

Despite the fact that numerous CAS exist, we will work mainly in the present paper with MAPLE (version 5) and DERIVE. Our main interest is in symbolic computation. In such a framework, the software becomes a kind of 'personal assistant' for computations and experimentations, for the researcher, as well as the teacher. One of the main problem with a CAS is how to interpret the results, in particular when the computations seem to have gone wrong. In this chapter, we will explore several situations[1] which will illustrate what we call *reasoned instrumentation*, in particular in a classroom context.

We will only suppose from the reader some basic knowledge of MAPLE or a similar CAS (e.g., MATHEMATICA, MuPAD or MAXIMA).

1.2 Technical preliminaries

It is useful to have a better understanding of how MAPLE (or equivalently your personal CAS) works. In our specific case, we can say that *Maple is written in... Maple.* Indeed, only 'basic' functions are built in another programming language (C in our case). This set of functions is called the *kernel* of MAPLE. All the other functions and procedures are written using functions from the kernel. This is a great advantage for learning. It is possible to see how most of the functions are written, thus giving us the possibility of modifying them. In order to do this, we use the command **interface (verboseproc=2);** described in Kofler (1997, Chapter 28). Then, it is enough to print the given procedure in order to see the code. In the following we give three examples:

```
> interface(verboseproc=2):
> print(abs);
                    proc() option builtin; 72 end
> print(goto);
                    proc() option builtin; 108 end
> print(sin);
proc(x::algebraic)
local n, t;
option
'Copyright (c) 1992 by the University of Waterloo. All rights
reserved.';
   if nargs <> 1 then ERROR('expecting 1 argument, got '.nargs)
   elif type(x, 'complex(float)') then evalf ('sin'(x))
```

```
  elif type(x, '*') and member(I, {op(x)}) then I*sinh(- I*x)
  elif type(x, 'complex(numeric)') then
     if csgn(x) < 0 then -sin(-x) else 'sin' (x) fi
  elif type(x, '*') and type(op(1, x), 'complex(numeric)') and
  csgn(op(1, x)) < 0 then -sin(-x)
  .
  .
  .
  elif type(x, 'function') and op(0, x) = 'JacobiAM' then
  JacobiSN(op(x))
  elif type(x, 'arctan(algebraic, algebraic)') then
     op(1, x)/sqrt(op(1, x)^2 + op(2, x)^2)
  else sin(x) := 'sin' (x)
  fi
end
>
```

The functions **abs** and **goto**[2] are part of the kernel, which is indicated by the **option builtin**. On the other hand, the function **sin** is a program of 60 lines and uses other functions. Another possibility in order to understand how a procedure works and what the procedure is doing is to trace it. In other words, we can trace all the different steps of the computation (assuming this functionality was allowed in the procedure). We have access to this functionality via the command **infolevel**. For instance, **infolevel [all]:=5**; will print the maximum of detail during the running of the procedure and this is a valuable tool for the teacher. For further information, see Kofler (1997: Chapters 28 to 30).

1.3 Representation of an object in a CAS

As aforementioned, we will illustrate this topic with MAPLE, but basically most CAS use similar features. In order to compare two expressions, we will need some normalization procedures, particularly in the difficult case where we want to know if a given expression is zero or not. Usually, a CAS like MAPLE does few evaluations of the expressions. But some systems, like DERIVE, perform more evaluations. We can illustrate that on a simple example; we have:

$$\sqrt{2}\sqrt{3} - \sqrt{6} = 0.$$

Suppose that we define the variable t, in MAPLE, as the expression:

$$t = \sqrt{2}\sqrt{3} - \sqrt{6}.$$

This last expression will return non-evaluated by the system and will not be replaced by 0. Nevertheless, the test **is(t=0)** will return *true*. In comparison, with DERIVE, we will have an immediate evaluation of the expression, with the expected result. In fact, with MAPLE, the variable t is stored as a symbolic expression (i.e., a syntactic expression). Thus[3], the quantity $\sqrt{2}\sqrt{3}-\sqrt{6}$ is nothing else than a sequence of symbols attached to the variable t. In this case, we have three components $op(0,t) = +$, $op(1,t) = \sqrt{2}\sqrt{3}$ and $op(2,t) = -\sqrt{6}$. Each one of these components could be split into other sub-components, in order to get, formally speaking, a tree structure. An example of normalization, under MAPLE, is given by the command **radnormal** (for *radical normal form*). If we apply this command to t, we get:

```
> radnormal(t);
                                    0
```

This procedure is quite efficient and could simplify (in an algorithmic way) any complex radical expressions. Some examples are given in the following:

```
> radnormal(sqrt(5+2*sqrt(6)));
                                  1/2    1/2
                                 3    + 2
> radnormal(sqrt9^(1/3)+6*3^(1/3)+9));
                                  1/3
                                 3    + 3
> radnormal((sqrt(3)+sqrt(2))/(sqrt(3)-sqrt(2)), rationalized);
                                      1/2  1/2
                                5 + 2 3    2
```

In the last command, we have added the option *rationalized* in order to normalize the denominators as well. The above operations are consistent with the following equalities:

$$\sqrt{5+2\sqrt{6}} = \sqrt{3}+\sqrt{2},$$
$$\sqrt{\sqrt[3]{9}+6\sqrt[3]{3}+9} = \sqrt[3]{3}+3,$$
$$\frac{\sqrt{3}+\sqrt{2}}{\sqrt{3}-\sqrt{2}} = 5+2\sqrt{6}.$$

In the same way, there is a *normal form* for polynomial and rational expressions. The uniqueness of such normal forms allows the system to compare such kinds of expression, and the efficiency of the internal representation of the expressions permits such comparison in reasonable time. In this type of algebraic treatment, CAS are very efficient and reliable. For more details related to radical expressions, we suggest reading Jeffrey & Rich (1999). In the following, we will give some insights on how the internal representation of the expressions is done, mainly by use of examples. Consider the expression below:

```
> E:=ln(1+x)-2*sin(1+x);
                              E :=ln(1 + X) - 2 sin(1 + x)
```

From a user point of view, the expression could be seen as a tree with, as root, the dyadic operator +. Using the commands **op** and **nops**, we can explore the apparent tree structure of the expression. In our example, we have:

```
> nops(E);
                                    2
> op(0,E);
                                    +
> op(1,E);
                                    ln(1 + x)
> op(2,E);
                                    -2 sin(1 + x)
```

Thus using a sequence of **op(op(op(...)))**, we can get a conceptual representation of the expression as below:

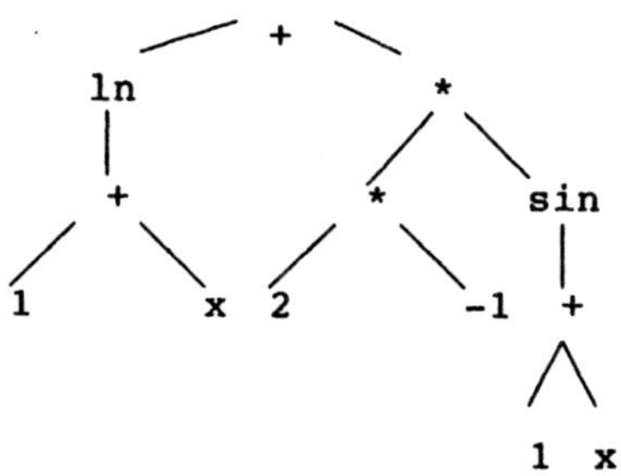

Nevertheless, this is not the true internal representation (i.e., in MAPLE) of the expression, which is in fact a little more complex but also more efficient. Technically speaking, the model for the internal representation is what is called a *DAG*[4], which allows the elimination of any common sub-expression (i.e., we only represent, and compute, once any given sub-expression which appears several times inside the main expression). It is difficult to see directly this representation in MAPLE. However, using the command **dismantle** (which we need to load with **readlib(dismantle)**), we can get a glimpse of the DAG structure. For instance, using our previous example, we get:

```
> dismantle[dec](E);
SUM(136478368,5)
```

```
  FUNCTION(136040888,3)
    NAME(135425704,4): ln #[protected]
    EXPSEQ(136609520,2)
      SUM(135549672,5)
        INTPOS(135407372,2): 1
        INTPOS(135407372,2): 1
        NAME(135426280,4): x
        INTPOS(135407372,2): 1
  INTPOS(135407372,2): 1
  FUNCTION(135443776,3)
    NAME(135426680,4):sin # [protected]
    EXPSEQ(136609520,2)
      SUM(135549672,5)
        INTPOS(135407372,2): 1
        INTPOS(135407372,2): 1
        NAME(135426280,4): x
        INTPOS(135407372,2): 1
  INTNEG(135407380,2): -2
```

The option **dec** allows us to identify the memory address (inside MAPLE) of the objects. For instance, an expression of the type **SUM(135549672,5)** means that we have a sum and its result will be put at the (memory) address 135549672. As we can see, the quantity 1 + x is put at this address and thus is computed only once. And so is ln(1 + x) (at the address 136040888). If another procedure uses this quantity, then it will point to this address. In order to understand this notion, we will digress slightly, and use a simple example:

```
> B:ln(1+x)^2;
                                                2
                                B := ln(1 + x)
> dismantle [dec](B);
PROD(136082120,3)
  FUNCTION(136040888,3)
    NAME(135425704,4): ln # [protected]
    EXPSEQ(136609520,2)
      SUM(135549672,5)
        INTPOS(135407372,2): 1
        INTPOS(135407372,2): 1
        NAME(135426280,4): x
        INTPOS(135407372,2): 1
  INTPOS(135407388,2): 2
```

No surprise there; the variable B 'points' to the expression ln(1 + x) previously defined in E. If we set x : = 3, since it is an integer we should have an immediate evaluation and B should be equal to $\ln(4)^2$; but doing so, B will point to a new address. Here is the session:

```
> addressof(B);
                                    136082120
> B;
                                           2
                                    ln(1 + x)
> addressof(x);
                                    135426280
> x:=3;
                                    x := 3
> addressof(x);
                                    135407404
> B;
                                          2
                                    ln(4)
> addressof(B);
                                    136154348
> E;
                                    ln(4) - 2 sin(4)
> dismantle [dec](B);
PROD(136154348,3)
   FUNCTION(136154236,3)
      NAME(135425704,4): ln # [protected]
      EXPSEQ(135824692,2)
         INTPOS(135407420,2): 4
   INTPOS(135407388,2): 2

> dismantle [dec] (E);

SUM(135484800,5)
   FUNCTION(136154236,3)
      NAME(135425704,4): ln # [protected]
      EXPSEQ(135824692,2)
         INTPOS(135407420,2): 4
   INTPOS(135407372,2): 1
   FUNCTION(136324224,3)
      NAME(135426680,4): sin # [protected]
      EXPSEQ(135824692,2)
         INTPOS(135407420,2): 4
   INTNEG(135407380,2): -2
```

We have used implicitly the command **addressof** in order to get the (memory) address of a MAPLE object (which we can see in the printing of **dismantle**). As we can see, the structure of the expression is left unchanged (B is still a sub-expression of E, and the DAGs are unchanged too), but the addresses have changed. As a matter of fact, MAPLE (as other CAS) uses an internal table for its objects, which maps each object to an address. When the value of an object changes, the table of addresses is updated, but losing the expressions already present in the memory. In our example, the address which corresponds to the expression ln(1 + x) is still there. Indeed, if we revert x to being a formal variable, we get our initial values:

```
> x :='x';
                              x := x
> E;
                   ln(1 + x) - 2 sin(1 + x)
> B;
                                      2
                           ln(1 + x)
> addressof(E);
                           136478368
> addressof(B);
                           136082120
```

This illustrates the value of the DAG representation and its efficiency. Its downside is the memory consumption. The reader will find further discussions of this topic in Giusti & al (2000) and at the end of Chapter 2 of Gomez & al (1995) and in Nizard (1997).

As a final point, we will show[5] how the DAG structure is suitable for computing formal derivation of an expression. Suppose we want to differentiate (formally) the symbolic expression:

$$g(x) = 2xe^{x^2}\sin(e^{x^2}).$$

Its DAG representation is given by:

$$\begin{aligned}
a_1 &= x,\\
a_2 &= a_1^2,\\
a_3 &= e^{a_2},\\
a_4 &= \sin(a_3),\\
a_5 &= a_3 a_4,\\
a_6 &= a_1 a_5,\\
a_7 &= 2,\\
a_8 &= a_7 a_6.
\end{aligned}$$

Then, we get the derivative of g by a formal derivation of each node of the DAG, resulting in a new DAG which is the associate to the derivative of g (say, with respect to x). If we denote by ∂ this formal derivation, then using the basic rules, we have:

$$\partial a_1 = 1,$$
$$\partial a_2 = 2a_1 \partial a_1,$$
$$\partial a_3 = e^{a_2} \partial a_2,$$
$$\partial a_4 = \cos(a_3)\partial a_3,$$
$$\partial a_5 = a_3 \partial a_4 + a_4 \partial a_3,$$
$$\partial a_6 = a_1 \partial a_5 + a_5 \partial a_1,$$
$$\partial a_7 = 0,$$
$$\partial a_8 = a_7 \partial a_6 + a_6 \partial a_7$$

We should notice the technical simplicity of this approach, which is independent of the choice of DAG structure.

Despite the fact that our data structure is efficient, it does not mean that we can always show that something which should be zero is *really* zero. In the same way, we cannot put all the 'special identities' in the CAS. For instance, in MAPLE, the expression $\sin^2 x + \cos^2 x = 1$ is not 'equal' to 1, even if x is assumed to be a real number. On the other hand, **simplify** 'knows' the identity:

```
> restart;
> a:=cos(x)^2+sin(x)^2;
                                    2          2
                      a := cos(x)  + sin(x)
> is(a=1);
                                FAIL
> eval(a);
                                 2          2
                         cos(x)  + sin (x)
> assume(a,real):eval(a);
                                  2           2
                        cos(x~)  + sin(x~)
> x :='x':simplify(a);
                                  1
```

But other (in fact most) well known identities cannot be reduced (even with **simplify**).

```
> restart;
> b:=arccos(x)+arcsin(x);
                   b :=arccos(x) + arcsin(x)
> simplify(b);
                     arccos(x) + arcsin(x)
```

But, we can use MAPLE to investigate this identity.

```
> diff(b,x);
                                0
```

```
> eval(subs(x=0,b));
                                    1/2 Pi
```

As a general fact, the procedures of normalization or *simplification* are rather limited. On one side, we cannot put all the 'special identities' in a computer, and on the other we often have multiple choices for the simplification (technically, speaking it is a *rewriting* problem). From a theoretical point of view, we have a kind of *decidability* limitation. Indeed, Richardson (1968) has shown that for the class $\mathcal{R}$ defined as the closure of rational functions of $\mathbb{Q}(\pi, \ln(2), x)$ under the action of sine, exponential and absolute value maps, the predicate $E = 0$ is 'recursively indecidable' for $E \in R$. Here the notion of *decidability* is relative to the theory of formal languages and grammars. A variation of this result to symbolic integration is given in Davenport & al (1987, p.168).

In practice, this is far less limiting than we might imagine. First, a lot of functions belong to a class for which decidability is established, and furthermore, there exist probabilistic methods which can be used efficiently in such settings. Nevertheless, this shows that a 'perfect' CAS, able to simplify 'ideally' any expression is a pure dream.

2. BASIC COMPUTATIONS AND LIMITS

While the main target users of a tool such as DERIVE are at high school level, this is quite different for MAPLE and other general CAS, which are designed for the engineer, the scientist and the researcher. Thus, such software is designed with generality in mind, and this can have some undesirable side effects for the student. We give a concrete example below:

```
> f:=x->evalf(x^(1/3));
                                       1/3
                         f :=x -> evalf(x   )
> f(-1);
                   .5000000000 + .8660254038 I
```

In general, and as seen in the classroom by the author, the student expects to get f(–1) = –1. But, the explanation is quite simple; it is enough to think complex! Indeed, if we search for a solution of the form $e^{i\theta}$, we reasonably get $\theta = \frac{\pi}{3}$, and it is this solution that the CAS gives. Most of the system operates, by default, on the complex numbers, as is also the case for advanced pocket calculators. Thus, we should show some caution when we use these tools, in particular in the classroom. Most of the time, problems can be solved by giving some help to the system. Here is a simple example: let *a* be a positive real number, and p an arbitrary real number. Set:

$$b = \log_a\left(\frac{a^{p^2}}{a^{p+1}}\right)$$

Taking account of the hypothesized characteristics of *a* and p, we should have $b = p^2 - p - 1$. But the expression is not simplified into this form by MAPLE, mainly due to the fact that, for the system, *a* and p are complex numbers. We need then to make our further assumptions clear to the system:

```
> b:=log[a]((a^p)^p/a^p/a);
                 (a^p)^p
             ln (------)
                  a^p a
        b := ----------
                ln(a)
> simplify(b);
             ln((a^p)^p a^(-p -1))
             ------------------
                   ln(a)
> assume(a>0);
> simplify(b);
             -ln(a~) + ln((a~^(p - 1))^p)
             ------------------------
                       ln(a~)
> assume(p,real);
> simplify(b);
                p~^2 - p~ -1
```

2.1 Computation of Limits

Taking limits of functions or of sequences is one of the most important notions of analysis. Hence, the behavior of CAS in this particular area is of prime interest. Several approaches have been developed to perform this kind of computation, and they are detailed in the works of Gruntz (1996, 1999). The first approach is what we can call the 'heuristic method'. In short, we use as main tool the Bernouilli de l'Hôpital rule[6] (which we will denote by BdH in the remainder of the text). The heuristic involves deciding how many iterations of the rule will be applied. Unfortunately, this method has several gaps. First of all, there are many types of expression which remain difficult to compute after applying the BdH rule, and furthermore, applying

BdH could increase the complexity of the expressions and thus the computations.

The second approach consists of using power series expansions (mainly Taylor expansions). This method is quite systematic, but it fails when:

- a zero expression is not detected by the system as being zero;
- there are an infinity (or at least a 'huge number') of zero terms in the power expansion. Such a situation appears when we compute the sum of functions, which could result in infinite cancellations;
- there are one or several parameters in the expression.

The first point, as already discussed, is very difficult in practice (Gruntz 1996, § 7.2; 7.3). A possible solution, depending of the type of expression, is to redefine the zero test used by the CAS. For instance, with MAPLE, it is done via the command *Testzero*, which by default uses the standard normalization procedures. An example is given later. Another possibility is to use some probabilistic method to perform the test.

The second point is well known to undergraduate students and is mainly related to the order of the power expansion; several methods exist to deal with this problem. The last point is more difficult in practice, and will mostly require use of heuristics.

To the best of our knowledge, DERIVE (and so the TI-92) uses a heuristic approach for the computation of limits, while MAPLE uses power series expansion methods. As we will test out below (and as we can see in the examples given in Wester 1999, p.166 and 168), CAS are quite fragile on limit computations. It is likely that the teacher (and the user) will need to take a lot of precautions, and in particular will need to know what to expect from the system. This is a typical situation where *reasoned instrumentation* is important.

We will now give several examples from a MAPLE session (in command line), that we will explain in detail at the end, and we will conclude by a discussion of the numerical approach to limit computations. Most of the following examples are extracted from or inspired by Wester (1999) and Gruntz (1996).

```
> # Example 1
> restart;
> # problem of the order
> limit((sin(tan(sin(x)^7))-tan(sin(sin(x)^7)))/x^49,x=0);
                                 7                   7
                sin(tan(sin(x) ))- tan(sin(sin(x) ))
          lim   ------------------------------------
        x -> 0                     49
                                  x
> Order;
                                   6
```

```
> Order:=50;
                              Order := 50
> limit((sin(tan(sin(x)^7))-tan(sin(sin(x)^7)))/x^49,x=0);
                                         -1
                                         --
                                         30
> # Example 2
> restart;
> A:=(sqrt(8)*(arccos(1/3+h)-arccos(1/3))/3/h)^2;
                           (arccos(1/3 + h) - arccos(1/3))^2
             A :=8/9 ---------------------------------
                                           h^2
> limit(A,h=0);
                                       infinity
> series(A,h=0,1);
8/9 (1/2 Pi - arcsin(1/3) - arccos(1/3))^2 h^-2 -
    4/3 (1/2 Pi - arcsin(1/3) - arccos(1/3)) 2^1/2 h^-1 +
    (-1/4 (1/2 Pi - arcsin(1/3) - arccos(1/3)) 2^1/2 +1) + O(h)
> # most of the terms in the above expression should be zero!
> # Example 3
> restart;
> A :=ln(h)/(ln(h)+sin(h));
                                              ln(h)
                               A := ---------------
                                         ln(h) + sin(h)
> limit(A,h=infinity);
                                            infinity
> series(A,h=infinity,2);
> Error, (in asympt) unable to compute series
> A1:=1/(1+sin(h)/ln(h));
                                              1
                               A1 := ----------
                                            sin(h)
                                        1 + ------
                                            ln(h)
> limit(A1,h=infinity);
                                            infinity
> limit(1+sin(h)/ln(h),h=infinity);
                                           1
> # this is in fact a problem of operations on the limit!
>
> # Example 4
> restart;
> C:=z/(sqrt(1+z)*sin(x)^2+sqrt(1-z)*cos(x)^2-1);
```

```
                                  z
          C := ------------------------------------------
                     1/2       2          1/2       2
               (1 + z)    sin(x)  + (1 - z)    cos(x)  - 1
> limit(C,z=0);
                                  0
> # wrong result !
> # trying to explain the behavior
> series(C,z=0,2);
                              1
                    ------------------- z + O(z^2)
                          2         2
                    sin(x)  + cos(x)  - 1
> # the coefficient in z is in fact infinity!
> # it is again a problem of testing equality to zero
> # Here is a way to solve the problem via some re-writing
> restart;
> Testzero:=t -> evalb(normal(convert(t,exp))=0);
           Testzero := t -> evalb(normal(convert(t, exp))=0)
> C:=z/(sqrt(1+z)*sin(x)^2+sqrt(1-z)*cos(x)^2-1);
                                  z
            C := -----------------------------------------
                     1/2       2         1/2       2
               (1 + z)    sin(x)  + (1 -z)    cos(x)  - 1
> limit(C,z=0);
                                  2
                          ---------------
                                2         2
                          sin(x)  - cos(x)
> # correct result!
> # Example 5
> restart;
> F:=(ln(x)^2*exp(sqrt(ln(x)*ln(ln(x))^2\
> )*exp(sqrt(ln(ln(x)))*ln(ln(x)))^3))/sqrt(x);
           2            1/2          2                  1/2               3
      ln(x)  exp(ln(x)    ln(ln(x)) ) exp(ln(ln(x))    ln(ln(ln(x))) )
   F:= ---------------------------------------------------------------
                                 1/2
                                x
> limit(F,x=infinity);
                                  0
> # which is also correct.
```

Comments

Example 1 illustrates the classic problem of 'guessing' in which order we need to perform the power series expansion in order to remove the indeterminate form. We should notice that computing the power series expansion at the start with a systematically large order is costly in terms of running time, and we also need to take this into account.

Example 2 is a 'test to zero' problem. The limit computed by MAPLE is incorrect. In order to understand what is going on, we do a power series expansion, and we notice that the method fails because the classical relation:

$$\frac{\pi}{2} = \arccos\frac{1}{3} + \arcsin\frac{1}{3},$$

is not know by the system (cf. previous section). In theory, the power series expansion should look like:

$$series(A, h = 0, 1) = 1 + O(h),$$

and we should conclude that the limit is 1 (correct answer). Both examples are correctly computed by DERIVE (TI-92), and as an exercise for the reader, we can check (with MAPLE for instance) that both limits can be performed with the BdH rule.

Example 3 is of a different type. Indeed, it is straightforward to show that the limit is equal to 1. Surprisingly enough, DERIVE gives the correct answer. But MAPLE fails due to the asymptotic expansion of the expression (it is the sine which is the culprit here). Transforming the expression does not help, despite the fact that MAPLE can compute easily that:

$$\lim_{h \to +\infty} \left(1 + \frac{\sin(h)}{\ln(h)}\right) = 1.$$

This is in fact a simple problem of operations on the limits. Since the limits of the numerator and denominator exist and are equal to 1, it should be possible to conclude that the limit of the expression is 1. Again, DERIVE computes this limit correctly, most likely by performing some basic operations on the limit.

Example 4 is of the same type as Example 2. The classical relation $\sin^2 x + \cos^2 x = 1$ is not known by the system and generates a singularity in the power series expansion. Here, the proposed solution is a modification of the "test to zero", by reprogramming the procedure *Testzero* with a more suitable normalization. Then, the computation can be done correctly. We should notice that we can get the limit directly via BdH, and thus DERIVE gets the correct answer immediately.

However, if in the expression C we replace x by 1/x and z by exp(–x), with $x \to +\infty$, then MAPLE gives a correct answer, namely –2, while DERIVE is unable to evaluate the expression.

Example 5 is straightforward for MAPLE, but is unsolved by DERIVE which remains in "Busy" position. It is in fact a typical example of a limit which is cumbersome for the heuristic approach. We can check that most of the limits (Wester 1999, p.166 and 168) are difficult for DERIVE.

2.2 Numerical limit vs. Symbolic limit

We might think that the computation of limits should be easily done via numerical evaluation of the function in the neighborhood of the point, which could be given by a numerical sequence of points in this neighborhood. This possibility is indeed implemented in MAPLE, and can be called by the command **evalf(Limit())**. For examples 2 and 1, we get:

```
> restart;
> A:=(sqrt(8)*(arccos(1/3+h)-arccos(1/3))/3/h)^2;
                                                       2
                   (arccos(1/3 + h) - arccos (1/3))
     A :=8/9 -----------------------------------
                                  2
                                 h
> evalf(Limit(A,h=0));
                            1.000000000
> evalf(Limit((sin(tan(sin(x)^7))- tan(sin(sin(x)^7)))/x^49,
x=0)); Error, (in evalf/limit/levinu) invalid cancellation of
infinity
```

As we can see, while we get a correct answer for Example 2, Example 1 is still problematic. In the same way, it is difficult to get a numerical estimate of the limit at infinity of a general function.

Furthermore, new problems could arise, as shown with the example below (Gruntz 1996, p.4-7).

```
> restart;
> C :=1/x^(ln(ln(ln(ln(1/x))))-1);
                                     1
                   C := --------------------------
                         (ln(ln(ln(ln(1/x)))) - 1)
                        x
> evalf(subs(x=10^(-3),C));
                       .00005602749469
> evalf(subs(x=10^(-6),C));
                                       -6
                       .6146316300 10
> evalf(subs(x=10^(-12),C));
                                       -9
                       .1528656745 10
> evalf(subs(x=10^(-24),C));
                                       -16
                       .7787481717 10
> evalf(subs(x=10^(-80),C));
                                       -39
                       .1401026662 10
> # Is the limit 0...?
> # 10^80 is quite close to the number of atoms in our
universe...
>
```

```
> limit(C,x=0,right);
                                        infinity
> # oops !
>
```

A plot of the function would confirm the 'numerical evidence', but the result given by the symbolic procedure is in fact correct (Exercise left to the reader). Even increasing the number of digits will not change the numerical evaluation. Indeed, for 1/x large (e.g., 10^{80}), the iteration of ln diminishes tremendously the effect of 1/x and thus the expression C behaves almost as $x^{1-\varepsilon}$ with 'ε small', which gives evidence for the 'limit 0'. But, in order to suppress the effect of the iteration of ln, we need to take $x < e^{-e^{e^{e}}}$, or say x smaller than $10^{-1656521}$...(sic!)

3. SYMBOLIC INTEGRATION

This is one area where the CAS is a genuinely valuable tool. Due to the ubiquity of integration, both in mathematics and the sciences in general, we often have to perform antiderivation (i.e., symbolic integration) in order to solve the problem. For several years, physicists and engineers had to use huge tables of antiderivatives, also well known to contain several errors. But CAS are particularly efficient in this task, due to the fact that we have reasonably efficient algorithms, which are consequences of mathematical results often deep and unknown to most mathematicians. The goal of this section is to describe the methods, and give some examples of *reasoned instrumentation*. We heartily suggest that the reader look at Davenport & al (1987, Chapter 5) for further investigations.

Usually, the hand computation of antiderivatives is strongly based on heuristics (inherited from personal knowledge) such as: use of tables, change of variables, integrating by parts, use of multidimensional integrals, and so on. As pointed out in Gomez & al (1995, Chapter 9, p.227): "These methods often allow us to find an antiderivative quite quickly, but if they fail you cannot conclude". The methods used by the CAS are both heuristic and algorithmic. The result depends strongly on the type of function that we are integrating. In particular, for a given class of functions, a failure of the procedure is in fact a *proof* that the antiderivative cannot be expressed in terms of 'elementary' functions. We will begin by giving definitions and explain the practical difficulties. We will mainly follow the approach of Davenport (1987, Chapter 5, § 5.1.3).

Definition 3.1

Let K be a field of functions over a field k (not necessarily a prime field). The function θ is said to be an elementary generator on K if one of the following conditions holds:

1. *the function θ is algebraic over K (i.e., satisfies a polynomial equation with coefficients in K);*
2. *the function θ is exponential over K (i.e., for a k-derivation D of K, there exists $\alpha \varepsilon K$ such that $D(\theta) = D(\alpha)\theta$);*
3. *the function θ is logarithmic over K (i.e., for a k-derivation D of K, there exists $\alpha \varepsilon K$ such that $D(\theta) = \frac{D(\alpha)}{\alpha}$).*

Remark 3.2

Despite the fact that we can take a '*k*-derivation' in a general sense (i.e., an element of $Der_k(K)$), in practice $K = k(t_1,\ldots,t_n)$ and $D = \frac{\partial}{\partial t_i}$ for a certain i, where the t_l are indeterminate. In order to get an elementary generator which is not algebraic over K, we add formally the element exp(u) with the rules of derivation of the exponential (idem for the log).

Now we can introduce the following notion.

Definition 3.3

Let K be a field of functions over the field k. A field $K(\theta_1,\ldots, \theta_n)$ containing K is called a field of elementary functions over K if each θ_i, $i = 1,\ldots,n$, is an elementary generator over K. A function is said to be elementary over K if it is an element of a field of elementary functions over K.

A result of Liouville, often called 'Liouville's principle', gives the general form of the primitive of an element of a function field K if the function is elementary over K. The crucial result which is used and its algorithm implemented in most CAS is the theorem of Risch (Davenport & al 1987, p.185).

Theorem 3.4 (Risch 1969)

Let $K = k(x, \theta_1,\ldots, \theta_n)$ be a field of functions over k. We suppose that each θ_{i+1} is either a logarithm, or the exponential of an element of $k(x, \theta_1,\ldots \theta_i)$, and is transcendental over this field. Then there exists an algorithm which, given an element f of K, either gives an elementary function on K which has f as derivative, or proves that f does not admit an elementary antiderivative on K.

This result illustrates the fact that a 'failure' gives non-trivial information about the problem. The proof of the theorem of Risch is understandable at the undergraduate level. To the best knowledge of the author, the algorithm of Risch implemented in MAPLE is only complete for the class of elementary functions which are purely transcendental (no algebraic dependency involved; for instance *arcsin* is not purely transcendental). Thus we can apply the theorem of Risch only to this class of functions. For further reading we suggest (Davenport 1987, § 5.1.8) and (Bronstein 1990, 1997).

In practice, particularly in the classroom, the main difficulty will be to understand that a simple modification of the expression to be integrated will result, most of the time, in an antiderivative quite different in appearance. We will give some examples below:

```
> diff(sqrt(4^x+1),x);
                              x
                             4  ln(4)
                       1/2 ----------
                                x 1/2
                           (1 + 4 )
> f1:=int(4^x/sqrt(1+4^x),x);
                               x 1/2
                          (1 + 4 )
                 f1 :=2 -------------
                             ln(4)
> # Here Maple identifies the derivative
> # Let us change 4^x to 2^(2*x)
>
> f2:=int(2^(2*x)/sqrt(1+4^x),x);
        infinity
         ------
        \
                                        k2
f2 := 2  )    GAMMA (_k2 + 1/2) (-1)-
        /
         ------
         _k2 = 0

/                                     /              ln(2)          \\ //
|1 - exp(- 1/2 x (ln(4) + 2 _k2 ln(4)) |1 - 4 -----------------|)|/|
\                                     \     ln(4) + 2 _k2 ln(4)// /\
                                      /              ln(2)          \\ /
   GAMMA(_k2 + 1 (ln(4) + 2 _k2 ln(4)) |1 - 4 -----------------|| /
                                      \     ln(4) + 2 _k2 ln(4)// /
     1/2
   Pi
> # oops !
> # Compare the two results ...
> evalf(series(f2,x));
                          2                    3                    4
.7071067812 x + .3675968038 x  + .09908837872 x  + .01103831202 x  +

                     5      6
   .00008501305392 x  + O(x )
> evalf(series(f1,x));
```

```
                                      2                  3
2.040278892 + .7071067810 x + .3675968036 x  + .09908837866 x  +
                 4                   5        6
   .01103831201 x  + .00008501305389 x  + O(x )
> Digits;
                                                   10
> Digits:=20;
                                          Digits := 20
> evalf(series(f1,x));
2.0402788931935789636 + .70710678118654752440 x + .3675968038007051968
     2                          3                          4
   9 x  + .099088378720063520381 x  + .011038312018296660770 x  +
                            5        6
   .000085013053929147665047 x  + O(x )
> evalf(series(f2,x)) ;
                                                    2
.70710678118654752440 x + .36759680380070519690 x  + .0990883787200635
        3                         4                              5
   20388x  + .011038312018296660770x  + .000085013053929147665054x  +
     6
   O(x )
>
```

In the above example, we can see the practical difficulty of antiderivative computations. We can check that the results are the same up to a constant (which is exactly $\frac{\sqrt{2}}{\ln(2)}$). In a classroom situation, without the help of the teacher, it is likely[7] that the students will claim failure of the software and will stop there. Here, the theorem of Risch says that we should anticipate a correct result. Nevertheless, we can try to do some manipulations, in order to see if we can avoid such difficulties:

```
> Q :=2^(2*x)/sqrt(1+4^x);
                                       (2 x)
                                      2
                                 Q := ---------
                                        x 1/2
                                     (1 + 4 )
> simplify(Q);
                                        x
                                       4
                                     ----------
                                        x 1/2
                                     (1 + 4 )
> QQ:= normal (Q, expanded);
                                       x 2
                                     (2 )
                                QQ := ---------
                                        x 1/2
                                     (1 + 4 )
> int(QQ,x);
                                   (2 x) 1/2
                              (1 + 2         )
                              ---------------
                                   ln(2)
```

As we can see, using **simplify** gives the original form of the expression. Furthermore, the normalization with the option **expanded**, results in an even simpler expression for the antiderivative. It could be wise to develop such an approach, by using **normal** for instance, in order to get a 'good' expression before integration.

Here are other illustrations of the theorem of Risch:

```
> int(1/ln(x^2+1),x);
```

$$\int \frac{1}{\ln(x^2+1)}\,dx$$

This shows, assuming no errors in the software or the hardware, that $\frac{1}{\ln(x^2+1)}$ has no elementary antiderivative, as this function is purely transcendental. But the situation is less simple, with non purely transcendental elementary functions. Here is an example from Gomez (1995, p.235-236):

```
> f3:=int(arcsin(x)^2,x);
```

$$f3 := \int \arcsin(x)^2\,dx$$

But this is a classic antiderivative. We need to help the system. For this purpose, we can do an integration by parts using the package **student**.

```
> intparts(f3,arcsin(x)^2);
```

$$\arcsin(x)^2\,x - \int 2\,\frac{\arcsin(x)\,x}{(1-x^2)^{1/2}}\,dx$$

```
> intparts(%,arcsin(x));
```

$$\arcsin(x)^2\,x + 2\arcsin(x)\,(1-x^2)^{1/2} + \int -2\,dx$$

```
> value(%);
                      2                       2 1/2
           arcsin(x)   x + 2 arcsin(x) (1 - x )     - 2 x
```

Hence, by interacting with the CAS, we get a correct answer. This indicates the right way to use a CAS, and it is what the student will have to learn.

3.1 Difference between antiderivative and indefinite integral

Another aspect, often neglected, is the study of the result of the computation of an antiderivation. Indeed, in the usual settings, we want to compute a definite integral of a function which in most cases will be continuous on the interval of integration. Thus, we expect the antiderivative of the function integrated to also be continuous (and even more). Unfortunately, in many cases, it is not. As we have seen in the previous section, the computation of the antiderivative is a purely algebraic process. We just ask to find an element g of a differential field such that D(g) = f for the derivation D of the field. On the other hand, in a classical course on integration, we introduce the notion of *indefinite integral* of a function (real or complex variable), which is often confused with the notion of antiderivative. To be more precise, we will give a definition of this notion:

Definition 3.5 (Jeffrey 1994, p.35)
An indefinite integral of a function f, Lebesgue-integrable on an interval I of $\mathbb{R}$, is any function G defined on I and a real number c, such that:

$$G(x) = \int_a^x f(t)dt + c\ .$$

One of the main results of undergraduate courses in calculus (Riemann rather than Lebesgue) is that the indefinite integral is an antiderivative of f. But if f is continuous on I, so is G. However, an antiderivative is not necessarily an indefinite integral, and the CAS gives us several examples:

```
> int(6/(5-3*cos(x)),x);
                        3 arctan(2 tan(1/2 x))
```

Here the result is an antiderivative on $\mathbb{R}$, but fails to be continuous. As a consequence, we cannot use this antiderivative 'out of the box', in order to compute the definite integral of the function on [0,4π] for instance. The result would be nonsense (but apparently most students are not afraid of that). Nowadays, new heuristics and algorithms try to produce better results, closer to an indefinite integral; Jeffrey (1994) for more details.

4. CONCLUSION

The main point that we have tried to establish, through several examples, is that we cannot reasonably use a CAS as a black box, in particular in the classroom. Teachers should have sufficient knowledge of the behavior of the CAS in order to understand the result for themselves, and also for their students. Working with a computer always involves a risk. But, while students will accept a system or hardware failure, they will be less inclined to accept failure from the teacher, which will undoubtedly undermine the often fragile trust between them. We should also take into account that a student will often use 'the' worst possible strategy in order to solve a problem; in general, this will require the teacher to improvise in order to explain either the failure or the success of the strategy. This provides further evidence that teachers need a sufficient knowledge of the CAS if they want to 'feel in control' in the computer classroom.

Furthermore, many teachers have generally accepted ideas about the CAS (and using CAS in the classroom) which could work against them. One of the most prominent ideas is the belief that working with a CAS means teaching (and training) the students how to compute (as in the classical first undergraduate courses on calculus and algebra). Unfortunately, as already seen in the examples discussed in this paper, this is not true. Indeed, students need to acquire enough practical experience to develop insight on the methods learned. For instance, this will help them to estimate the order needed to compute a limit using power series expansions of composed functions, or the transformations to perform in order to compute an antiderivative. On the other hand, computing the power series expansions of functions up to a high order as the climax of a calculus course is pointless.

This is why we believe that a *reasoned instrumentation* should be preferred when we have to use CAS in the classroom. The 'intelligent usage' of a CAS in mathematical science courses (or in other sciences) is not an easy task, and in particular is not given. This implies the necessity of developing specific classroom activities and specific exercise sheets, well adapted to the task, showing clearly the value of the CAS either as a platform for experimentation or as an assistant, and well integrated into the main course.

Philippe ELBAZ-VINCENT (pev@math.univ-montp2.fr) is an assistant professor at the Université Montpellier II (France). His research area is algebra and number theory, in which he develops computational methods.

NOTES

1. The examples given have been produced with the version V release 5.1 of MAPLE under GNU/LINUX-2.4 (architecture i686) and SOLARIS 5.8 (architecture Ultrasparc II) and with the DERIVE version implemented on the TI-92 pocket calculator (Texas Instruments).
2. This last function is undocumented...
3. We can access the components of this expression via the command **op**, and **nops** to get the number of components (as coded by MAPLE).
4. Directed Acyclic Graph.
5. Based on an example from Ole Stauning for the software FADBAD of automatic formal derivation.
6. The famous rule of the Marquis de l'Hôpital (1661-1704) is in fact due to Johan Bernoulli, 1667-1748, cf. (http://www-groups.dcs.st-andrews.ac.uk/~history/Mathematicians/Bernoulli_Johann.html).
7. As experienced several times by the author.

REFERENCES

Bronstein M. (1990) On the integration of elementary functions, *Journal of Symbolic Computation* 9(2), 117-173.

Bronstein M. (1997) Symbolic Integration I - Transcendental Functions, *Algorithms and Computations in Mathematics* (Vol. 1). Heidelberg: Springer-Verlag.

Crandall R.E., Mayer E.W. & Papadopoulos J.S. (1999) *The twenty-fourth Fermat number is composite*, preprint.

Davenport J., Siret Y. & Tournier E. (1987) *Calcul formel: systèmes et algorithmes de manipulation algébrique*. Paris: Masson.

Dumas P. & Gourdon X. (1997) *Maple: son bon usage en mathématiques*. Berlin: Springer-Verlag.

Fortin P. *Introduction au calcul formel*. Paris: Texas Instruments.

Giusti M., Hägele K., Lecerf G., Marchand J. & Salvy B. (2000) Computing the Dimension of a Projective Variety: the Projective Nœther Maple Package, *Journal of Symbolic Computation* 30(3), 291-307.

Gomez C. Salvy B. & Zimmermann P. (1995) *Calcul formel: mode d'emploi, exemple en Maple*. Paris: Masson.

Gruntz D. (1996) *On computing Limits in a Symbolic Manipulation System*, Ph.D Thesis ETHZ.

Gruntz D. (1999) Computing Limits in Computer Algebras Systems, in M.J. Wester (Ed.), *Computer Algebra Systems: A practical guide* (pp. 153-170). Wiley.

Jeffrey J.D. (1994) Integration to obtain expressions valid on domains of maximum extent, in M. Bronstein (Ed.), *Proceeding of ISSAC 93* (pp. 34-41). ACM Press.

Jeffrey J.D. & Rich A.D. (1999) Simplifying Square Roots of Square Roots by Denesting, in M.J. Wester (Ed.), *Computer Algebra Systems: A practical guide* (pp. 61-72). Wiley.

Kamerich E. (1998) *A Guide to Maple*. New York: Springer-Verlag.

Kofler M. (1997) *Maple: An Introduction and Reference*. Addison Wesley.

Lazarus A.J. (1994) Implications of Computational Mathematics, in W. Gautschi (Ed.), *Mathematics of Computation 1943-1993: A Half-Century of Computational Mathematics, Proceeding of Symposia in Applied Mathematics* 48 (pp. 561-564). AMS.

Nizard A.F. (1997) *Structure des objets Maple*, exposé stage de Luminy, option Info des classes de prépa, CIRM (http://pauillac.inria.fr/~quercia/documents-info/Luminy-97/nizard/articl97.ps).

Plestsch B. (1999) Computer Algebra in Mathematics Education, in M.J. Wester (Ed.), *Computer Algebra Systems: A practical guide* (pp. 285-322). Wiley.

Richardson D. (1968) Some Undecidable Problems Involving Elementary Functions of a Real Variable, *The Journal of Symbolic Logic* 33(4), 414-520.

Trouche L. (1998) Expérimenter et prouver: faire des mathématiques au lycée avec des calculatrices symboliques. Montpellier: IREM, Université Montpellier II.

Von zur Gathen J. & Gerhard J. (1999) *Modern Computer Algebra*. Cambridge University.

Wester M.J. (Ed.) (1999) *Computer Algebra Systems: A practical guide*. Wiley.

Chapter 3

TRANSPOSING COMPUTER TOOLS FROM THE MATHEMATICAL SCIENCES INTO TEACHING

Some possible obstacles

Jean-Baptiste Lagrange
IUFM de Reims & DIDIREM, Université Paris VII, France
jb.lagrange@reims.iufm.fr

Abstract: Thanks to the work of mathematicians, software designers and mathematics educators, computer algebra is now available on calculators that students can afford for classroom use. These new artifacts certainly open up stimulating prospects, but we should not look on them just as a miraculous solution to difficulties of teaching. We ought to initiate an in-depth reflection on their educational use in relation to the wider evolution of mathematics.

In this chapter, we will discuss how the new tools offered to students fit into the evolution of mathematics itself, and of mathematics teaching and learning. We will also consider difficulties in adapting teaching which often make integrating these new tools something of an adventure.

Key words: Algorithms, Experimentation, Mathematical Sciences, Transposition.

1. THE FUTURE OF MATHEMATICS TEACHING

In recent years there has been much discussion in France about the future of mathematics teaching (Chapter 1). An official committee, the CREM (Chapter 1, § 2.3; 4) was created to think about what this future should be. According to this committee, a dramatic change has occurred over the past 50 years; mathematics is now produced and used by a great variety of people. One consequence is that we have to change our conception of mathematics to consider *the mathematical sciences* which are not just the concern of mathematicians.

The impact of ICT on mathematics teaching was a central issue in the discussion. The CREM committee devoted a report to this topic. According to this report, as ICT pervades all of society, mathematics is to be found everywhere in modern life, but people have no means to perceive this. Thus mathematics teaching should aim to make people aware of this hidden mathematics. ICT has also changed mathematics itself as the mathematical sciences now include an *experimental dimension* helped by the wide use of mathematical software.

According to the CREM, the use of mathematical software could help mathematics education to adapt to this new situation. A drawback is the sophistication of the mathematical software proposed for students' use, contrasting with the conceptual simplicity and clarity of traditional mathematics teaching. According to the CREM, the possibility of achieving a similar simplicity and clarity should be secured by teaching basic ideas of data processing.

Box 3-1.

Extract from the CREM report

(Kahane 2002)

In the recent evolution of mathematics, many new sources and outputs have appeared, as well as considerable work on existing mathematics. Other sciences and practices have provided mathematics with new problems, methods and concepts. New concepts and theories have been created and have sometimes proved useful in unexpected applications. Simulations based on mathematical models are present in every scientific activity and the development of mathematics benefits from both internal and external interaction. Thus mathematics is far from being just a matter for mathematicians. Contemporary mathematics can be described as a pumping, distilling and irrigating process involving physicists, computer scientists, engineers, biologists and economists together with mathematicians in the strict sense of the word.

As the CREM noted, curricula have recognized a need for such teaching but never really implemented it because of a lack of curricular contents and classroom activities. According to the CREM new prospects are opened up by the use of mathematical software, provided that teaching can clarify for students how mathematical data is represented and processed. Thus, basic notions about data representation and processing should be studied, together with numerical and symbolic experimental processes, number representation and processing, induction, graphs, convexity, continued fractions... This could provide the framework for a new algorithmic way of thinking in the common culture.

2. INTEGRATING NEW TOOLS: A PROBLEM OF TRANSPOSITION

The goal of this book -- conceptualizing the integration of symbolic calculators -- is consistent with the above concern. The numerical, graphic, symbolic *and* programming capabilities of the new calculators to be found at school level are a transposition of computerized tools used in mathematical sciences. Thus they can play a major role in the future of mathematics teaching, not just as pedagogical aids but as a vehicle for new approaches. However, their use could conceal the mathematical basis of these approaches if teaching does not adapt its goals, contents and methods.

In this chapter, we will take advantage of didactical theorization and classroom observation to better understand these new approaches, and to look into the way in which teaching could be adapted. First we will focus on the global question of the impact of the mathematical sciences on teaching, taking into account that the purposes of professional mathematicians and researchers differ from the aims of teaching. We will use the notion of *didactical transposition*. When he introduced this notion Chevallard (1985, p.14) stressed that "what happens inside a didactical system cannot be understood without considering what happens outside[1]". He proposed to consider "genesis, filiations, gaps and reorganizations" interrelating mathematics teaching and professional mathematics. In this approach, mathematics in research and in school can be seen as a set of knowledge and practices in transposition between two institutions, the first one aiming at the production of knowledge and the other at its *study* (Chapter 5, § 1). French curricula clearly consider this prospect when they stress the role of the calculator in helping experimental approaches and the use of spreadsheets or calculators in carrying out algorithms (Box 1-1, Box 3-2). Filiations clearly appear but there must be hidden obstacles because this prospect was never really implemented in classrooms.

We see a cause of these obstacles in the different aims of the two institutions. Professional mathematics favors new approaches and reorganizations on the basis of productivity and mathematical correctness. In official research fields, data and methods to obtain conjectures using mathematical software are now being published and discussed. Some mathematicians specialize in the production and publication of experimental outcomes, while other mathematicians use these conjectures to work on proofs.

> Experimental mathematics is that branch of mathematics that concerns itself ultimately with codification and transmission of insights within the mathematical community through the use of experimental exploration of conjectures and more informal beliefs and a careful analysis of the data acquired in this pursuit (Borwein & al 1996).

In addition, computer science techniques in research motivate fundamental mathematical work about algorithms. Effectiveness -- existence of an algorithm to solve a given problem --, complexity -- the algorithm's properties in relation to processing time and data size -- and efficiency -- the practical conditions for the use of the algorithm on a given technology -- are important notions in this work (Rouiller & Roy 2001, p.35).

In contrast, teaching, especially in the general (rather than vocational) stream, is not primarily interested in improving mathematical productivity by way of new tools but rather in the transmission of a mathematical culture (Chapter 9, § 1). The kernel of this culture lies in the social expectations of parents, students and teachers, and generally does not change easily. In order to survive in contemporary societies where ICT has a major role, this kernel should now integrate the potential of new tools and mathematical activity inspired by the mathematical sciences. Because the kernel was built when only traditional tools existed, this integration has a cost -- a not-obvious in-depth reorganization -- and resistance can be expected. To look at this reorganization, we will distinguish two dimensions, one about algorithms and the other about experimental approaches. These dimensions are certainly closely interrelated in the mathematical sciences, but, as we shall see, obstacles to their integration into teaching and learning are dissimilar.

3. ALGORITHMS IN MATHEMATICS TEACHING

Curricula have recommended the use of programmable calculators for fifteen years. A study of textbooks and practices shows that only graphic and numerical capabilities have actually been used in classrooms. More recently, the Euclidean algorithm for the greatest common divisor appeared at the 9th grade and prime number search algorithms were introduced in the 12th grade. To look into the difficulties of transposition we will consider the 9th grade curriculum, analyzing how an algorithmic idea was transposed from advanced mathematics and how it was (mis)understood in textbooks.

The curriculum (Box 3-2) introduces the Euclidean algorithm in terms of two dimensions. The first one is theoretical: a fundamental property of divisors helps to *build* the algorithm. The second dimension is practical: offering the algorithm as a means to recognize and obtain irreducible fractions. In addition, this dimension helps to *insert* the algorithmic approach inside the 'usual' mathematics. From the study of these two dimensions, mathematical and algorithmic meaning can be expected.

Box 3-2.

Algorithmic approach in the French grade 9 curriculum

"The goal is to develop an overview of numbers and to emphasize algorithmic treatments"

Content	Intended proficiency	Comments
... 4- Integer and rational numbers. Common divisors. Irreducible fractions.	Being able to find whether two given numbers have a common divisor greater than one. ... Being able to simplify a given fraction into an irreducible fraction.	...The sum and the difference of two multiples of an integer are themselves multiples of this integer. It is then possible to build an algorithm (Euclid's or another). This algorithm will give the GCD of two integers and answer the question [of knowing whether a fraction is irreducible]... Teaching will take advantage of spreadsheets and CAS for this topic.

This introduction is consistent with the approach to algorithms current in advanced mathematics. For instance, the extracts from a computer science classic (Box 3-3) show that, at this level, the Euclidean algorithm helps to understand the notion of data processing, and to introduce a method for the formulation of algorithms. On one hand, the idea of transposing this approach into secondary teaching is stimulating because the algorithm appears to derive logically from properties of the GCD and from mathematically expressed constraints of execution (decrease of $x + y$). Emphasizing the links between an algorithm and the underlying mathematical properties is certainly something interesting to transpose into secondary education. On the other hand, the method deals with difficult logical concepts like *weakest precondition* and *weakest condition such that the execution is guaranteed to decrease a function.*

Box 3-3.

Euclid's algorithm in a computer science classic
(Dijkstra 1976)

Chapter 0: Executional abstraction

Let us consider a mechanism. On a cardboard with grid points, the only numbers written are the values 1 through 500 along both axes. The 'answer line' with the equation x = y is drawn. When we wish to find the greatest common divisor of two values X and Y, we place a pebble on the grid point with the coordinates x = X and y = Y. As long as the pebble is not lying on the 'answer line', we consider the smallest equilateral rectangular triangle with its right angle coinciding with the pebble and one sharp angle on one of the axes. The pebble is then moved to the grid point coinciding with the other sharp angle of the triangle. The above move is repeated as long as the pebble has not yet arrived on the answer line.

Chapter 7: Euclid's algorithm revisited

... I shall now devote yet another chapter to Euclid's algorithm. I expect that in the meantime some of my readers will already have coded it in the form

```
(x,y):= (X,Y);
do x>y→x:=x-y;    y>x→y:=y−x od;
print(x)
```

Let us now try to forget the cardboard game and let us try to invent Euclid's algorithm afresh (...). Collecting our knowledge we can write down:

GCD(X,Y)=GCD(Y,X)

GCD(X,Y)=GCD(X,X+Y)=GCD(X,X-Y)

GCD(X,Y)=abs(X) if X=Y

...This is strongly suggestive of an algorithm that establishes the truth of P=(GCD(X,Y)=GCD(x,y) and x>0 and y>0)... whereafter we 'massage' (x,y) in such way that the relation P is kept invariant. If we can manage this massaging process so as to reach a stage satisfying *x* = *y*, then we have found our answer by taking the absolute value of x (...).

For the assignment x:=x-y we find the weakest condition such that the execution is guaranteed to decrease x+y is y>0, a condition that is implied by P.

Full of hope, we investigate the weakest precondition in order that P is valid after the assignment x:=x−y. (We find) GCD(X,Y)=GCD(x-y,y*)* and x−y>0 and y>0. The outmost terms can be dropped as they are implied by P and we are left with the middle one.

Thus we find x>y→ x:=x−y and x>y→ x:=x−y...

Box 3-4.

Euclid's algorithm in a 9th grade textbook
(Chapiron & al 1999)

A method using iterated subtractions helps to find the Greatest Common Divisor of two numbers. This method is very old and known as Euclid's algorithm. The flow chart [below] explains how it works. Calculations are easy but repetitive and sometimes long. A spreadsheet helps to calculate more quickly.

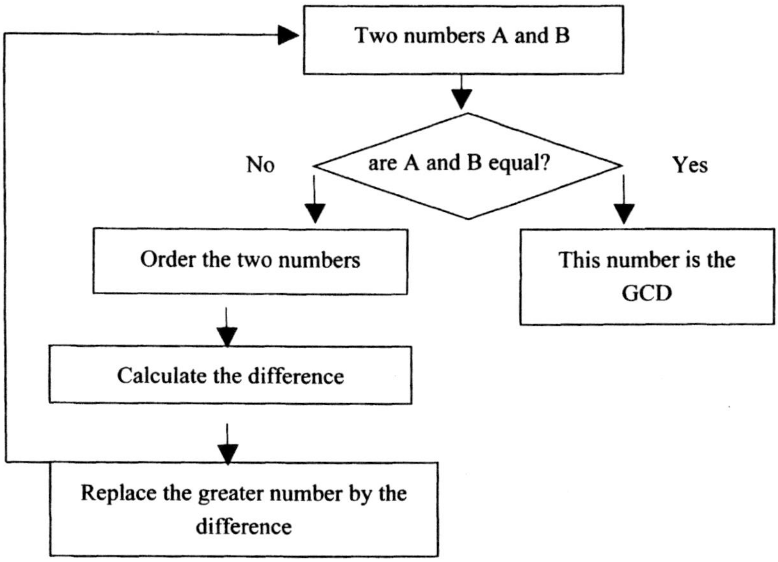

It appears that the curriculum considered the interest of the method but not the difficulties of its existence as an isolated object. Proof of this can be found in the way in which textbooks (mis)interpret the curriculum. In contrast with the curriculum and Dijkstra's book, a typical textbook (Box 3-4) presents neither practical nor theoretical dimensions of Euclid's algorithm. It offers just a flow chart and a 'push button' translation into a spreadsheet. Other textbooks also adopt this approach. The reason why textbooks did not follow the curriculum when it offered the algorithm for practical simplification of fractions is that they judged that this use is not realistic. The logical difficulties of constructing the algorithm from properties of the GCD and the constraints of execution certainly explain why textbooks offer no theoretical dimension. Obviously, the curriculum tried to introduce Euclid's algorithm without much change in the more general background. Under these conditions the algorithm is an isolated object without a real mathematical existence, and textbooks are not able to make something interesting of this object.

Introducing a real algorithmic dimension by using calculators in mathematics teaching and learning would imply greater change. But, then another difficulty would be that there is very little didactical research able to offer help in thinking about this change. Research about the learning of algorithmic processing tends to be old and isolated. Rogalski & Samurçay (1990) found important conceptual difficulties, even with very simple iterative processes, but no other research studies followed. There are also too few research studies about data representation (Aharoni 2000 is a very isolated example). The case of the algorithmic approach shows that, even when powerful tools are available, we remain a long way short of achieving a cultural algorithmic way of thinking, through a satisfactory transposition of advanced mathematics to teaching and learning in the secondary school.

4. EXPERIMENTAL APPROACHES IN TEACHING AND LEARNING

As compared to algorithms, the experimental dimension of mathematical sciences seems at first sight more easy to transpose into the kernel of mathematical culture, as mentioned above. In mathematical research, producing and trying conjectures helps to discover new theorems and to build new theories. Experimental approaches cannot have this role in teaching, but other contributions are generally expected. Authors like Pérez Fernández (1998) and Kutzler (1997) start from the idea that experimenting can help students to develop a more in-depth understanding of mathematics. They generally stress that 'traditional teaching' does not work properly because students have just to repeat routines and are not allowed to search by themselves. In these authors' view, experimenting with new tools will be a remedy. For instance, students, even with weak abilities in arithmetic or algebraic procedures, might be able to use symbolic calculators to explore advanced mathematical domains or to try several approaches to problems that they could not do by hand. Thus, with new tools, mathematical teaching would become more interesting and accessible to more students.

On one hand, using the potentialities of new tools in experimental approaches is a stimulating idea. The visualizing capabilities of computers could help more varied access to mathematical problems and concepts. Exploring a problem, students could study a number of examples, using varied representations of objects and inductive as well as deductive approaches. On the other hand, in learning, as in mathematical sciences, understanding of concepts does not emerge spontaneously from observation, even with the help of powerful tools.

Experimental approaches in research are harnessed to the production of new knowledge. This articulation is what Borwein & al (ibid., p.16) named "theoretical experimentation" and includes structuring a domain to formulate hypotheses, deriving examples to try with a machine, interpreting the machine results... Transposition should thus maintain the linking of experimentation with theoretical elaboration. Reflecting on this transposition at the beginning of the TI-92 experiment, we found that, in teaching and learning, this articulation is far from obvious, and that authors stressing the potentialities of new tools generally tended to underestimate the difficulties of their classroom use. To explain this, let us look at classroom situations involving an experimental activity.

We consider two classes of situations. In the first class, students have to observe and interpret a number of calculator answers. For instance, approaching multiple representations of a concept, students have to consider perceptually how a mathematical property appears as a phenomenon in several windows of a calculator; or in an applied mathematics course, students are meant to use a CAS to avoid too complicated hand calculation. In the second class of situations students have to experiment on symbolic phenomena and find general structures by induction. In other words, students have to observe 'how the machine does it', try to discover techniques 'to do the same as the machine' and give a mathematical interpretation of the machine operation. The two classes of situations differ in terms of the consideration given to the machine operation. In the first class, a student is expected just to use the computer output as mathematically relevant results, whereas in the second s/he is meant to think about how they were obtained.

Berry & al (1994) identified "five potential ways to integrate the use of DERIVE into a mathematical course" (Box 3-5). The first class of situation that we consider in this chapter corresponds to potential way 2 and 5 ("as a problem solving assistant" and "an aid to visualization and interpretation") and the second class corresponds to potential way 3 ("as an investigative or exploratory environment").

Let us consider a situation from the first class, related to the idea of function. To help students to approach this idea, it is important to provide them with varied views on the relationship between representations. For instance one view of the relationship between algebraic definition and graphic representation arises through considering that a zero of the function corresponds to an intersection point between the graphic representation and the x-axis. It is not a spontaneous view because, at first, students see the intersection's coordinates as the solution of a two equation system. To help students to interpret graphically the zeros of a function, it is interesting to ask them to consider together these values -- obtained by algebraic solution -- and the coordinates of the intersection points read from the graph. Using symbolic calculators is helpful because the two operations are performed by specific commands in two separate windows.

Box 3-5.

Classroom situations involving an experimental activity

(Berry & al 1994)

DERIVE as a problem solving assistant (p.84)

In many mathematical modeling courses greater emphasis is placed on the formulation of problems and the interpretation of results, rather than the solution of the mathematical problems that may occur. Similarly, courses in applied mathematics may wish to concentrate more on the concepts and relationships that form the basis of the study but may also include extensive use of algebraic manipulation or calculus. In the past the emphasis in these types of courses has become distorted due to the large amount of time spent by the students obtaining solutions to problems compared to the important formulation and interpretation stages (...). By reducing the time that students need to spend obtaining solutions and increasing their reliability and their accuracy, DERIVE allows more emphasis to be placed on the formulation and interpretation phases of mathematical modeling or applied mathematics. In particular it allows students to use mathematical concepts and techniques that they understand in principle, can apply in simple cases, but find difficult to apply in the more complex cases that may arise in real problem solving.

DERIVE as an aid to visualization and interpretation (p.93)

It is important for students to be able to visualize and interpret mathematical results. Often with some higher level mathematics it is quite hard for students to do so and DERIVE can offer an environment in which it is possible to do mathematics and create visual images that allow students to interpret and comment on the results they have obtained. A student may show that a Maclaurin series approximation to $\sin x$ is $x - x^3/6$. It is quite hard however for the student to relate this back to the original function, or establish the range of values over which the approximation is reasonable. It is very simple to produce a series approximation of this type on DERIVE and compare its plot with a plot of the original function.

DERIVE as an investigative or exploratory environment (p.86)

It is possible to use DERIVE as an environment in which students can exploit and learn new mathematics by making discoveries for themselves (...). One very real benefit of this approach is that the students can gain an intuitive feel for mathematical ideas and principles before they receive a formal introduction to the mathematics. As an example consider the chain and product rules for differentiation. A typical text book introduction would give either a formal statement or derivation of the rule followed by worked examples. Alternative approaches designed to develop in students an initial intuitive feel for these rules have been laborious for students to carry out. DERIVE however offers the potential to develop these intuitive ideas very easily.

Situations of this type are often presented to promote the use of new tools, but it is seldom mentioned that they are effective only if students have a suitable preparation. Wain (1994) reports on an observation of students not able to recognize the decimal value that they read in the graphic window as an approximation of the symbolic solution that they obtained in the algebraic window. In contrast with the teacher's expectation, they could not see a relationship between zeros and intersection points.

Our opinion is that, before experiencing this situation, students should have been prepared to recognize the differing form taken by a number under several types of expression, particularly exact symbolic expressions in the symbolic window and approximate values in the tabular or graphic window. Guin & Delgoulet (1997) designed and experimented with classroom activities to achieve this preparation (Chapter 9, § 2).

Box 3-6 provides another example of a situation where it was assumed that CAS use would be transparent to students. DERIVE was supposed to help students to acquire better methods of transforming trigonometric expressions by performing automatically the more technical part of the transformation. But it failed because of students' misunderstanding of what a simplification process involves for a machine. Ruthven's analysis put to the fore the 'sense of the command' that would be necessary. Even when CAS is 'just a tool' an understanding of its technical operation cannot be avoided.

In the second class of classroom situations, symbolic capabilities are means for students to carry out algebraic transformations before knowing how to perform paper-and-pencil techniques and even before knowing the existence of these techniques and their mathematical underpinning. After a first encounter with notions like limits or derivatives, students could use a symbolic calculator to experiment with symbolic transformations (limits or derivatives of sums and products, for instance) and become aware of algebraic rules (or algebraic techniques) applied by the machine. They could then imagine general structures underpinning these rules.

This use of computer tools is also a stimulating prospect but we have to be aware that for students, detecting symbolic phenomena and inductively formulating algebraic rules and techniques might not be so obvious. During the TI-92 calculator experiment (Chapter 9, § 3) we had to reflect on how an experimental inductive activity about symbolic techniques for limits and derivatives could really be made to work.

Our first concern was to find questions that could provoke students' inductive reflection. Questions like 'observe what happens' do not necessarily yield interesting observations. Many algebraic transformations actually maintain the structure of most expressions -- for instance when the sum of limits is the limit of the sum -- and such examples are, of course, not problematic. Even when the structure is not maintained -- for instance in the case of indefinite limits or of differentiation of a product -- students do not spontaneously start thinking inductively. Results that the teacher expected to be amazing to students (for instance DERIVE simplifications for the chain and product rules of differentiation in Box 3-5) did not alone create much surprise. The learning situation has to bring to the fore puzzling peculiarities and to challenge students' anticipations. For instance it is often interesting to prepare examples where students' predictions will very probably be wrong and to ask them to compare these predictions with the calculator's answer.

Box 3-6.

Difficulties with emergent subgoals. The task of transforming the trigonometric sum, sin(x)+sin(2x), into a trigonometric product

(Ruthven 2002, p.287-288)

The French research provides a specific example in an episode in which students -- relatively experienced in using CAS -- were charged with the task of transforming the trigonometric sum, $\sin(x)+\sin(2x)$ into a trigonometric product (Lagrange 2000, p.5).

In response to a programmed command, the CAS gave the expression $2\sin((2x+x)/2).\cos((2x-x)/2)$. The students wanted to simplify this to $2\sin(3x/2).\cos(x/2)$. To their surprise, the response of the CAS to repeated simplify commands was first to give the original expression and then $\sin x+2\sin x.\cos x$. To the students, aware of the overarching goal, the emergent subgoal of simplifying the subsidiary algebraic expressions $2x+x$ and $2x-x$ was clear. To them, this was transparently the sense of the command to simplify. In other words, their articulation of the simplification operation was a situated one. But, of course, no model of the larger task -- and no situated sense of the command -- was available to the CAS; it was unable to take account of the wider mathematical context giving rise to the instruction. The machine is unable to interpret or adapt an instruction to accord with the wider purpose so evident to a user; it can only operate literally, either in terms of the formal elections made by the user, or of preset defaults, which, as in this case, may fail to coincide serendipitously with the wider purpose of the user. The effective instrumentation of mathematical reasoning by means of a CAS depends, then, on precise reframing of situated purposive actions into the decontextualised formal register of the machine, and a corresponding reframing of results.

A second concern was the knowledge that students need in order to understand observations as phenomena within an inductive approach, to construct interesting examples, and to interpret the calculator's answers correctly. An example from Pozzi (1994) will help in examining this question. Pozzi reports on an observation of two students who were trying to find a general rule for differentiating a product by observing how DERIVE computes derivatives of the product of a polynomial with a trigonometric function. Asking DERIVE to differentiate $\cos x(7x^3+2x)$, they got

$(21x^2+2)\cos x - x(7x^2+2)\sin x$. They then concentrated on the central part, $\cos x - x(7x^2+2)$, which they found very similar to the initial expression. They tried to induce a general rule involving the transformation of a product into a difference. Of course the central part has no meaning because it is not a sub-expression of the derivative. But, to students, it appeared to be the key to finding a rule because it is perceptually close to the initial expression. Students' algebraic knowledge about the structure of expressions was not strong enough to counterbalance this perceptual evidence and they could not make good use of DERIVE's help.

Thus, considering new tools as providing 'scaffolding' (Kutzler 1997) for weak student knowledge is an idea which needs to be re-examined. From Pozzi's example it is clear that we have to reflect on the prior algebraic knowledge required. Students do not necessarily need strong procedural abilities but obviously should not be lacking some key knowledge of algebraic structure (Chapter 7). Finally we had another concern about knowledge of the target concept itself in relation to the machine operation. For instance if students follow an experimental approach to a concept like limit mainly using the transformational capabilities of CAS, they will then associate the concept too closely with transformational rules and/or develop a 'push button' conception of the concept. Other modifications to the meaning of concepts may result from computer implementation and also interfere with experimental computer aided activity (Chapter 5).

5. CONCLUSION

This chapter started from the idea of transposing approaches from the mathematical sciences into teaching and learning as a major avenue through which to make sense of the use of new computer tools. We have seen that this is not so easy. In particular, a real teaching of algorithms is not feasible today because the traditional cultural kernel underpinning curricula is resistant, and too few research studies and experiments have been undertaken. Transposition of experimental approaches seems more viable, but difficulties are very often underestimated. We located obstacles to computer aided experimental activity making an intended contribution to conceptualization. It appears that using computer symbolic tools as resources in perceptual and inductive approaches requires reflection on what prior knowledge students need both in algebra and about the machine, on what questions can serve to provoke inductive thinking, and on what form students' representation of concepts and of the machine operation takes.

The consequence is that experimental computer-aided approaches to teaching and learning cannot be thought of as simply a matter of using a machine to ease problem solving or to enhance inductive activity. Following Ruthven (2002), using CAS for graphic and symbolic reasoning "influences the range and form of the *tasks and techniques*[2] experienced by students", and because tasks and techniques are resources available for more explicit codification it also influences "the theorization of such reasoning". Chapter 4 will give an example where teachers participating in a common project take radically different and somehow restricting options regarding management of this phenomenon in the classroom. Chapter 5 will provide further insight both practical and theoretical. Task and technique will be offered as structural levels organizing the study of a mathematical domain and connecting experimental approaches to graphic and symbolic problem solving with conceptual elaboration.

Jean-Baptiste Lagrange is a professor at the Institut Universitaire de Formation des Maîtres in Reims.

He has contributed to the development of CAS use in schools and to the associated didactical reflection. His present work includes the design of a CAS environment for teaching and learning at secondary-school level and investigation of the classroom implementation of this environment. He is also working to contribute to a better understanding of the professional situation of teachers trying to integrate ICT.

NOTES

1. Our translation.
2. Our emphasis.

REFERENCES

Aharoni D. (2000) What you see is what you get - the influence of visualization on the perception of data structures, *Proceedings of PME 24* (Vol. 2, pp. 1-8). Hiroshima: University.

Berry J., Graham T. & Watkins A. (1994) Integrating the DERIVE program into the teaching of Mathematics, *The International DERIVE Journal* 1(1), 83-96.

Borwein J., Borwein P., Girgensohn R. & Parnes S. (1996) Making sense of Experimental Mathematics, *The Mathematical Intelligencer* 18(4), 12-17.

Chapiron G., Mante M., Mulet, R. & Perotin C. (1999) *Mathématiques*, Collection Triangle. Paris: Hatier.

Chevallard Y. (1985) *La transposition didactique*. Grenoble: La Pensée Sauvage éditions.

Dijkstra E.W. (1976) *A discipline of programming*. Upper Saddle River: Prentice Hall.

Guin D. & Delgoulet J. (1997) *Etude des modes d'appropriation de calculatrices graphiques et symboliques dans une classe de Seconde*. Montpellier: IREM, Université Montpellier II.

Kahane J. (2002) *L'enseignement des sciences mathématiques*, Centre National de Documentation Pédagogique. Paris: Odile Jacob.

Kutzler B. (1997) With the TI-92 Towards Computer Age Maths Teaching, *International Journal of Computer Algebra in Mathematics Education* 4(1), 7-32.

Lagrange J.B. (2000) L'intégration d'instruments informatiques dans l'enseignement: une approche par les techniques, *Educational Studies in Mathematics* 43(1), 1-30.

Pérez Fernández J. (1998) Los systemas de cálculo simbólico en la enseñanza de las matemáticas, in Hodgson et al (Eds.), *Selected lectures 8th International Congress on Mathematical Education* (pp. 345-368). SAEM THALES.

Pozzi S. (1994) Algebraic Reasoning and CAS: Freeing Students from Syntax? In H. Heugl & B. Kutzler (Eds.), *DERIVE in Education* (pp. 171-188). Bromley: Chartwell-Bratt.

Rogalski J. & Samurcay R. (1990) Acquisition of Programming Knowledge and Skills, in J.M. Hoc, T.R.G. Green, R. Samurçay & D.J. Gilmore (Eds.), *Psychology of Programming* (pp. 157-170). London: Academic Press.

Rouiller F. & Roy M.F (2001) Calculer avec des inégalités. Introduction aux algorithmes de la géométrie algébrique réelle, in J.B. Lagrange, D. Lenne (Eds.), *Calcul formel et apprentissage des mathématiques* (pp. 35-47). Paris: INRP.

Ruthven K. (2002) Instrumenting mathematical activity: Reflections on key studies of the Educational use of computer algebra systems, *International Journal of Computers for Mathematical Learning* 7, 275-291.

Wain G. (1994) Some Technical Problems in the Use of DERIVE with School Pupils, *International DERIVE Journal* 1(1), 49-56.

Chapter 4

THE INFLUENCE OF A COMPUTER ALGEBRA ENVIRONMENT ON TEACHERS' PRACTICE

Margaret Kendal*, Kaye Stacey* and Robyn Pierce**
** The University of Melbourne, Australia, ** The University of Ballarat, Australia*
kendal@bacchusmarsh.net.au, k.stacey@unimelb.edu.au and r.pierce@ballarat.edu.au

Abstract: Using a computer algebra system (CAS) in the classroom provides many opportunities for improving student learning. However, to take advantage of such a powerful instrument as CAS requires changes to many aspects of the classroom. The different ways in which three pioneering Australian teachers adapted their teaching to use CAS are described (see also Appendix 4-1, for a comparison with similar experiences in other countries). Two of the teachers taught an eight-week course in each of two consecutive years (the CAS-Calculus project) at secondary school level, using a symbolic calculator. They gave CAS different roles in the instruction and in defining their curriculum goals. One teacher used CAS in a restricted way with the primary goal of increasing understanding while the second teacher adopted CAS as another technique freely available for solving standard problems and emphasized efficient routines. Over several years (in a separate project, at university level, using the computer program DERIVE), the third teacher has evolved a method of teaching with CAS, moving from an early emphasis on teaching about CAS as a tool and using it for difficult problems to incorporating its use for primarily pedagogical aims. In reporting on these case studies we comment on different ways of organising the classroom; the variety in approaches to teaching the use of CAS; the increased range of methods for solving problems and for teaching; the contrast between using graphics and symbolic calculators; the place of paper-and-pencil skills; devoting time to mathematics or technology; and finally the curriculum and assessment changes required in schools.

Key words: Assessment, Curriculum, Functional use of CAS, Graphics calculators, Pedagogical use of CAS, Symbolic calculators, Teaching styles.

1. INTRODUCTION

The technological tool, CAS, with its powerful symbolic, graphical, and numerical capabilities is becoming increasingly available to students of mathematics (Chapter 1). Calculators with CAS are highly suitable for use in secondary school classrooms since they are portable, can be used by students whenever they do mathematics, and are becoming more affordable. As yet, only a few countries around the world have established national policies in regard to CAS use in schools. It is interesting to compare the differences in priorities, curriculum, and assessment regimes that have emerged. This chapter describes the experiences of three teachers who have been pioneering the use of CAS in Australia (see Appendix 4-1 for other examples). Some of the findings and constraints reflect these teachers' individual contexts whilst others are universal.

Many Australian teachers have substantial experience in teaching with graphics calculators and are now becoming interested in exploring the use of the symbolic features of CAS on computers and calculators. Australian people are generally keen to try new ideas and to adopt new technology such as video-cassette recorders, mobile phones, and internet banking and there is a general feeling within the education system that students should be encouraged to use new technology. This is especially so in Victoria, the Australian state where these studies took place. This preference is moderated to some extent by a strong feeling for equity so that the introduction of a new technology may be delayed because not all students can afford to buy it. In the case of CAS, it is also moderated by an uncertainty about the effect of powerful technology use on basic calculus[1] and algebra skills, a concern that underpins the approaches of two of our three teachers. As is the case in France (Chapter 1), teaching mathematics in the senior secondary years in Australia is largely determined by the external examinations that students undertake at the end of school. There is currently a new senior mathematics course (Victorian Curriculum and Assessment Authority 2003) that permits students to use CAS in their examinations. However, Andre and Benoit, two of the research project teachers whose early work teaching with CAS is described in this chapter, were operating with a former curriculum that did not allow CAS in examinations.

This chapter begins with describing the teaching and learning in three classrooms. We highlight the ways in which the teachers have changed to accommodate CAS in their classrooms and the benefits and challenges that they believe it has brought. Benoit and Andre were teaching in a secondary girls' school and they volunteered to participate in a two-year research project conducted by the University of Melbourne. During the CAS Calculus[2] project, they taught their Year 11 students (most are 16 years old) introductory calculus using symbolic calculators (TI-92). They no prior experience in teaching with CAS but they and their students had used graphics calculators routinely in their other work. All of the lessons of both teachers, jointly prepared with the researchers, were observed and Andre and Benoit were interviewed before and after the project. Claire[3], the third teacher, has taught at a university incorporating the use of CAS in the computer program (DERIVE) for over eight years and more recently with symbolic calculators (TI-89). Since her students generally had limited school mathematics backgrounds the curriculum she adopted revisited much of the senior school mathematical content. However, she was free to organise the assessment as she decided (within constraints set by the university). The information in this chapter comes from her own reflections on her teaching, from the data gathered from her students by questionnaires and interviews, from analysis of students' work, and from analysis of curriculum materials including assigned exercises and examinations.

The experiences of these three pioneers serve to pinpoint a wide range of issues that emerge when teachers begin teaching with new technology.

2. BENOIT: TEACHING FOR UNDERSTANDING

In 1998, the University of Melbourne research team initiated the CAS-Calculus project to investigate the use of CAS for the teaching of introductory calculus. Over an eight-week period, we monitored three secondary school teachers, Andre and Benoit (from the same school) and Charles (from a different school) and their classes. They had no prior experience using a symbolic calculator or teaching wirh it. Together, the research team and the teachers planned the lessons, aiming primarily to develop students' conceptual foundation for differentiation, especially by use of multiple representations: linking graphs, symbols, and tables of function data. Results of the 1998 study are fully reported, by Kendal & Stacey (1999) and McCrae & al (1999), including a description of the three teachers' pedagogy and how each teacher's privileging impacted on student learning outcomes. In 1999, the study was repeated and Andre and Benoit participated again, with their new classes in the same school. The changes that occurred in Andre's and Benoit's pedagogy in the second year are reported in Kendal & Stacey (2001, 2002).

A central feature of the initial planning for the project was designing a course which emphasized multiple representations and where understanding would receive the major emphasis by preceding procedural work, as done by Heid (1988). CAS was initially designed by mathematicians to increase their efficiency in problem solving involving algorithms (Chapter 3). Hence, it presents as an ideal tool to relieve students of the tedium of paper-and-pencil computations and algebraic procedures. It dispenses with much of the routine symbolic manipulation and enables larger, more realistic problems to be solved. While traditional courses have largely emphasized procedures at the expense of understanding of related concepts, CAS-supported courses may enable students to access concepts that previously required prior skill in computation. Studies by Heid (1988), Palmiter (1991) and Repo (1994) showed that CAS could provide support so that early mathematical experiences can be concept-based. Lagrange develops this point further (Chapter 5).

Benoit was a very experienced teacher of senior mathematics and was Head of the Mathematics Department in his school. He was very interested in teaching with the graphics calculator (officially endorsed for all forms of assessment including state-wide examinations at the time) and he actively encouraged other teachers to use it in their classes. Benoit had attended professional training about teaching with graphics calculators and he was keen that his students were aware of all of their calculators' capabilities and could use them efficiently. One indicator of his interest and expertise is that he collected a wide range of programs and then downloaded them onto the students' calculators for their use in examinations.

When Benoit volunteered for this experiment, introducing the symbolic calculators, he was especially interested in using them to give students a better conceptual understanding of calculus. Benoit always emphasized understanding the concepts being taught. He frequently used enactive representations (e.g., making purposeful hand and arm movements in the air) and visualization techniques to explain symbolic ideas. For example, he constantly linked the symbolic derivative to gradient of the tangent to the curve (represented by his outstretched arms). He also used real world phenomena to explain mathematical ideas. For example, after discussion with the research team, he explained the rule for finding the derivative of a sum of two functions by considering the speed of a person running on a moving platform. On another occasion, to discuss a problem about maximising volumes, he constructed a box out of cardboard.

Benoit's teaching style was student-centred. He involved every student in the class by challenging them to explain their ideas to him and to the other members of the class. He overtly encouraged students to construct their own meaning for mathematical ideas through conjecture, analysis, negotiation of meaning with other students in the class, making decisions, drawing conclusions, and demonstrating conjectures. The construction of his blackboard notes was basically guided by his lesson plan but he also spontaneously incorporated key aspects of the class discussion. Benoit moved around the classroom and checked individual students' work as they solved problems from worksheets (or the textbook) that were developed to work on in class and complete at home. He responded to common difficulties by initiating further class discussion that helped the students to resolve their specific problems.

Benoit also used orchestrated discussion to teach his students how to use the symbolic calculators. He did not use an overhead projector or any other special classroom arrangement to demonstrate CAS procedures (Chapter 8); he would say what he was doing on his calculator, slowly enough so that all the students could follow on their own machines. Benoit would wait until the whole group reached each stage and walk around the classroom helping students who were in trouble. In this way, he ensured that all the students could follow on their own machines. Because of his expertise in group management, everyone participated. He only used an overhead projector for special demonstrations such as a dynamic experiment involving the collection of real data by a *data-logging device* attached to the calculator.

Benoit had an over-riding concern for developing students' understanding of the concept of differentiation. This was displayed in his strong emphasis on the links between functions and their graphs and between the derivative and the slope of tangents to curves. To Benoit, knowing these links was the essential aspect of understanding differentiation. A few years earlier he had embraced graphics calculators wholeheartedly because they had provided excellent technological support for this. The monitoring by the research team showed that this aspect of his teaching was reflected in what his students learned. The tests showed that most of his students were able to interpret a derivative in terms of the slope of a tangent or as a rate of change. In providing this focus he often used the graphical facility of the symbolic calculators.

On the other hand, Benoit's concern for understanding led him to restrict the use of the symbolic facility of the calculator. He was pleased to use it to perform repetitive routine tasks quickly as a preliminary step to discovering patterns and developing algorithms. For him the priority was developing the understanding of concepts through exploration, investigation, and induction:

> Potentially, it enables you to do a bit more investigation, in terms of looking at more complex functions... It's good for discovery, and it's a lot easier in terms of discovering [mathematical properties] because it takes a lot of hack work out of [it].

Benoit reported, for example, that through orchestrated discussion, the class had constructed tables of derivative values of polynomial functions and deduced the rules for the differentiation of x^n, ax^n and sums of these. He commented:

> I think we've done very nicely with the calculator. One thing I like is the routine procedures. You haven't got all that time wasting. You can do very nicely a lot of the

algebra. You can do it so simply on the calculator and you're avoiding in some ways the time that goes by when you're doing a lot of repetitive calculations.

Beyond uses such as this, Benoit carefully controlled when and how students used their symbolic calculators. He believed that doing algebra by paper-and-pencil methods was extremely important for understanding and that if he allowed his students to do algebra constantly with CAS he would be depriving them of an opportunity to understand. He stated in an interview that there are *"certain [algebraic] skills that kids have to have, even if you can use the technology... They've still got to have hands-on, they've still got to get pen-and-paper skills"*.

There were two changes to Benoit's privileging of, and preferences for, representations of differentiation and technology in his second year of teaching with CAS. One change was that Benoit reduced the use of CAS for symbolic differentiation, replacing it by an increased emphasis on students' performing algebra and differentiation calculations using paper-and-pencil. He did this, at least in part, because he assessed this particular class to be less mathematically able and in need of more practice at paper-and-pencil techniques. Another change was that Benoit omitted all work with the numerical (tabular) representation of differentiation, since he believed that this less able class would be confused by a third representation. He continued to emphasize the graphical representation of derivative but he omitted finding approximate or exact derivatives from tables of function values.

Benoit's decision-making about the emphasis to be placed on paper-and-pencil skills therefore seems to have been affected by three factors. Firstly, he was influenced by the amount of paper-and-pencil practice he believed the class required. Secondly, he noted that performing algebra step-by-step with paper-and-pencil contributes to a sense of 'understanding', an important issue that is addressed in Chapter 5. Thirdly, Benoit was always mindful of the fact that these students could not use CAS in their examinations. Although use of the symbolic calculator was permitted during the experimental CAS-Calculus teaching program and for the associated testing, Benoit would certainly have looked to the different requirements of the state-wide examinations 15 months ahead. Interestingly, amongst the graphics calculator programs that he provided for his students to take into the examinations was one that factorised quadratic expressions symbolically. This indicates that Benoit used technology to advantage his students' grades and raises the likelihood that if symbolic calculators had been available in examinations, Benoit may have made different decisions.

3. ANDRE: TEACHING FOR PERFORMANCE

Andre, a colleague of Benoit and also an experienced teacher of senior mathematics, taught a parallel class the same introductory calculus material. Unlike Benoit, Andre did not enjoy teaching with the graphics calculator. He recalled his previous experience in an interview:

> Actually, I tried to bring in the [graphics calculator] but I had real trouble with it. I thought *"I just can't be bothered"* and I haven't [used it in class] since. I didn't feel comfortable with the [graphics calculator] because I had so many problems.

In contrast, Andre enjoyed using the symbolic capabilities of the TI-92 both for personal use and in his class. As the project progressed he began to use it for demonstrating all procedures while teaching. He stated:

> I hooked it up at the beginning of the lesson and I used it much more than I would use a graphics calculator in the classroom... I was in the mode of always having it there and having it set up so that the overhead projector was there and I just slipped on the screen, hooked it up and it was there... It was on all the time and I felt comfortable.

Andre's normal teaching style was to lecture, emphasizing rules and procedures, with few teacher-student or student-student interactions. The students mostly worked alone but occasionally consulted with their neighbours. They were given the same mathematical questions as Benoit's students to work on in class and complete at home. During the research project, Andre used an overhead projector to demonstrate how to use the symbolic calculator to achieve given results. His students copied down two sets of notes as he wrote them up on the blackboard: first, the paper-and-pencil procedures; and second, the corresponding set of step-by-step symbolic calculator procedures. An example of Andre's blackboard notes is given in Figure 4-1. In contrast, Benoit gave much less emphasis to the procedures for using the calculator itself, and he generally managed to teach students how to carry out the procedures by demonstrating them to the whole class and then assisting individuals.

The number of bacteria (*N* millions) present in a culture at time *t* (in seconds) is

$$N = 10t^3 - 10t + 4$$

Find the rate of growth of the number of bacteria after 5 seconds.

With paper-and-pencil:	*With TI-92:*
$N = 10t^3 - 10t + 40$	[2nd] [8]
$\frac{dN}{dt} = 30t^2 - 10$	↓ $d(10x^3 - 10x + 40, x)$
$\frac{dN}{dt} = 30(5)^2 - 10$	[2nd] [K]
$= 750 - 10$	↓ $d(10x^3 - 10x + 40, x)\|x = 5$
$= 740$ million	[E] 740

Figure 4-1. Andre's blackboard notes showing both paper-and-pencil and TI-92 (symbolic calculator) procedures to solve a rate problem

Andre and Benoit had very different attitudes to student use of the symbolic calculator. Benoit controlled the use that students made of it, especially its symbolic facility, suggesting when students should and should not use it. In contrast, Andre gave his students complete freedom about when they could use their calculators. After his presentation to the class, his students were free to work using either paper-and-pencil techniques or symbolic calculator methods using the relevant step-by-step guides to procedures they had written in their notes. These guides also usually included the use of compressed calculator commands to carry out routine procedures in a minimum number of steps. So, for example, Andre encouraged his students to use one-line procedures such as:

Solve (*d*(x(32 – 2x), x) = 0, x)

which both differentiates the expression x(32 – 2x) and solves to find the point where the derivative is zero (Figure 4-2).

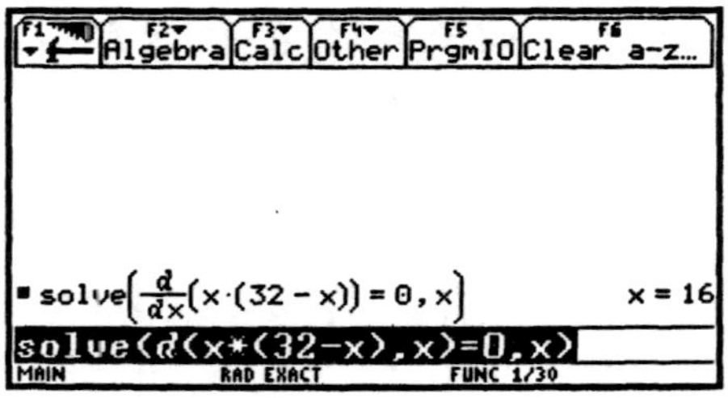

Figure 4-2. Result of a compressed use of TI-92

Whereas Benoit privileged the symbolic and graphical representations of derivative and the link between them, Andre had a strong preference for the symbolic representation. This seemed to be because it led to exact answers. In an interview before the second teaching trial, when asked to discuss alternative methods of finding the gradient of the curve at a point given the function and its graph, he indicated this dislike of ambiguity:

> Oh well, [differentiation and substitution] is accurate. An approximate method... would be actually drawing a tangent at x = 1 and then working out the gradient. But I don't think the girls[4] would ever attempt that because they hate anything where they have to guess or where the answer might be really different. But that doesn't necessarily mean that they would use the exact method like mine.

Further evidence of the importance of exactness to Andre came during the second trial, when he realized that the project evaluation would test differentiation in different representations. Some questions, for example, would require students to find the derivative of a function from the equation of the tangent line and other questions would require them to find approximate values of a derivative from a table of values. From this time on, in addition to the symbolic facility, Andre began to encourage the students to use alternative calculator methods to find derivatives. In particular, he encouraged them to explicitly calculate the numerical difference quotient $(f(x + h) - f(x))/h$ for a small value of h and a given value of x. He also encouraged them to use the calculator to draw a tangent at the specified point and read its gradient from the equation displayed on the screen. Andre was motivated to do this by his belief that the answers obtained by these methods were exact (in fact, the answers are not exact, but for simple functions and whole number coefficients they usually coincide with the exact values). Thus, during the trial, Andre increased his use of the symbolic calculator to include additional graphical and numerical differentiation procedures with CAS. Since he really liked using the symbolic calculator, he experimented with the data logger: several times he enthusiastically demonstrated (using projection) a dynamic program that linked the numerical and graphical representations of derivative. Both of these were new initiatives for him.

4. DIFFERENT TEACHER PRIVILEGING AND DIFFERENT RESULTS

4.1 More options for solving problems

The advent of new technological tools such as CAS is accompanied by an increased number of ways of solving mathematical problems. Methods that in the past were extremely tedious and so were only available *in principle* are now available *in practice.*

For example, it has always been possible *in principle* to solve equations by graphing, supported by mathematical analysis to predict the number of solutions and in what general regions they might be located. However, solving equations graphically used to be a method of last resort, when other methods failed, not a method of first choice. With advanced graphing capabilities on calculators and computers, this is no longer the case: solving equations graphically can be quick and easy.

A second example of the different status of mathematical methods with improved calculation concerns differentiation. Before scientific calculators, differentiation of the square root function might have been used to estimate $\sqrt{100.2}$ rather than calculate it numerically. This could be achieved by using the approximation of $f(x + h)$ by $f(x)+f'(x).h$ with $x = 100$ and $h = 0.2$.

Now we could use the reverse situation to find quickly the derivative (0.05) of the square root function at 100 with the calculation:

$$\frac{\sqrt{100.2}-\sqrt{100}}{0.2} \approx 0.049975$$

Alternatively, the function can be graphed and 'zooming in' can locate an approximate value, or automatically drawing the tangent also gives the derivative (Figure 4-3).

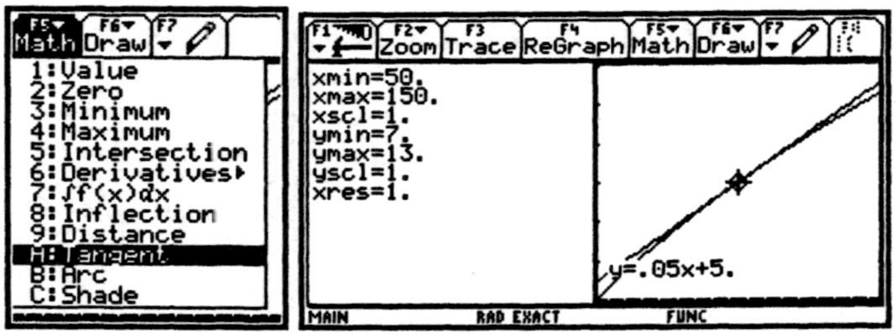

Figure 4-3. On the left, the command *Tangent* in the TI-92; on the right, the result for the square root function at the point x = 100

The National Council for Educational Technology in the United Kingdom acknowledged this explosion of feasible methods of solution in the new technological environments:

> For any one problem there may now be a range of methods of solution. Typically, there may be numerical and graphical approaches as well as algebraic and analytic approaches. Indeed there may be a variety of algebraic approaches. Hence it is more likely that a problem will be tackled with a view to comparing and contrasting different methods, with each solution possibly giving rise to some new mathematics (NCET 1994).

A consequence of having a greater choice of methods was that Andre and Benoit taught different ways of solving differentiation problems. Benoit taught his students to work primarily from the symbolic derivative, calculated using paper-and-pencil methods, and interpreted as the gradient of the tangent to a curve. In contrast, Andre's students had a wider range of methods for calculating derivatives (at a point) since they might work symbolically, or calculate a difference quotient from a table of values, or get the calculator to draw a tangent to the graph and then read off its gradient. We expect that this explosion in methods will be the norm: differentiation is not special in this regard.

4.2 More options for teaching

The growth in options for solving problems with new technologies is accompanied by a growth in options for teaching. Although Andre and Benoit worked together and taught similar students in the same school and planned the lessons together with the research team, the lessons that they gave were distinctly different. However, each adopted teaching practices with technology that supported their beliefs about mathematics and teaching mathematics. The two teachers had very different conceptions of mathematics: their teaching styles, use of representations (numerical, graphical, or symbolic), and general use of technology were distinctly different. In consequence, although the two classes had similar overall achievement, the students learnt rather different mathematics (for a more detailed analysis see Kendal & Stacey 2001, Kendal 2002).

Andre demonstrated use of the technology and provided clear flowcharts for students to follow in order to carry out routines efficiently. He was attracted to CAS because it could give accurate answers (or in one case, approximate answers that Andre believed to be accurate). It enabled Andre to extend his teaching and his students' skills with a new set of routine procedures for using CAS that matched his usual lecture/demonstration style of teaching, for teaching rules. He really appreciated the symbolic capability of CAS and enjoyed using it.

Benoit, in contrast, privileged pedagogical use of CAS. He saw this pedagogical use residing in two possibilities to increase understanding. Firstly, he believed that linking the symbolic and graphical representations of a function (or a derivative function) was a key to understanding. Secondly, the CAS enabled students readily to collect symbolic data for class discussions during which students would induce the rules for differentiation and so forth. Beyond use of this nature, the symbolic manipulation facility of the calculator was of little interest to Benoit.

4.3 Teaching with a graphics or symbolic calculator

Interesting differences were observed between the ways the teachers taught with a graphics calculator and with a symbolic calculator. Andre preferred the symbolic to the graphics calculator while Benoit preferred the graphics to the symbolic calculator.

These preferences are consistent with observations by other researchers. Teachers who view the graphics calculator as a tool for computation tend to stress content-orientated goals: for them, pupils learn by listening to the teacher's instruction. Tharp & al (1997) noticed that rule based teachers (like Andre) subscribed to the view that graphics calculators may hinder elementary learning: they restrained students' use of the graphics calculator. Andre did not like using the graphics calculator since he was expected to use it in ways (such as experimentation and discovery) that conflicted with his rule-based conception of mathematics and his preferred lecture/demonstration style of teaching during which he controlled the learning. In contrast, Andre was very enthusiastic about using the symbolic calculator:

> I loved it [symbolic calculator]. I thought it was great. I really liked the exact and approximate [i.e. the modes which specify whether answers are given as decimals or as rational numbers or roots], the spreadsheets [tables facility] and graphing from the spreadsheets. Yes, I thought they were fantastic. And the girls did too. I pined for it when I went back to the graphics calculator which I found very limited and inaccurate. [Some comments on the usefulness of the larger screen and better menu structure of the symbolic calculator followed].

Benoit's reactions to the two types of calculators were the opposite of Andre's. As indicated earlier, Benoit was highly skilled in using graphics calculators personally and he enjoyed teaching with then. He used then to help students understand concepts, particularly in explaining symbolic ideas graphically, and for explorations. His behaviour was consistent with other research, which suggests that teachers (like Benoit) who tend to employ interactive or inquiry-orientated methodologies use graphics calculators more than others. This other research also shows that teachers who see the graphics calculator as a tool for learning are more likely to have student-centred goals, interactive inquiry driven teaching styles, and student-centred views in learning. Tharp & al (1997) noticed that the teachers who were not rule-based teachers did not restrict student use of the graphics calculator for investigations and were more likely to be concerned with student conceptual understanding and thinking.

However, Benoit taught differently with a symbolic calculator. While he was willing to use it to perform repetitive symbolic tasks to generate patterns quickly and accurately prior to inductive reasoning, he was extremely wary of using its other symbolic capabilities, for reasons outlined above. In consequence, in both studies, he restricted students' use of the calculator.

5. CLAIRE: FROM FUNCTIONAL TO PEDAGOGICAL USE

In this section, we describe the journey of Claire as she experimented with CAS in her teaching. The university where Claire teaches first incorporated CAS using computer programs (DERIVE and MAPLE) into the mainstream mathematics and engineering mathematics courses in 1990 (as part of a technology-based project). At first, CAS was used for occasional demonstrations during lectures. Students used it in isolated sessions scheduled in the computer laboratories, where an important purpose was to familiarize students with the use of a potential workplace tool. In 1995, problem-based projects involving the use of CAS were introduced. These projects integrated mathematics and engineering through problem analysis, mathematical modelling, simulation, and solution validation. In the light of these experiences and the emerging research literature, CAS was more fully integrated into both the teaching and assessment of the mainstream mathematics courses from 1997. Pierce & Stacey (2001a) report the gradual changes in curriculum, teaching approaches, teaching materials, and assessment that occurred. During 2000, the teachers began to introduce hand-held symbolic calculators into their teaching.

Claire, who has been part of this transition since the mid-nineties, has documented her experiences. Up until 1997, she taught a traditional mathematics curriculum with CAS 'added on'. This involved teaching rules and procedures (as prescribed by the university) with an emphasis on paper-and-pencil methods. To aid visualization, CAS demonstrations were interspersed with the lectures. In laboratory sessions, the students completed worksheets that focused on how to use DERIVE to quickly and accurately perform the difficult calculations and complex symbolic manipulations. As a result of these experiences Claire realized that CAS could be used pedagogically (to assist students to *learn* mathematics) not only functionally (to *do* mathematics).

In 1997, a new first year unit was introduced to provide students who had limited secondary mathematics with the necessary pre-requisites to undertake mainstream university mathematics. Claire was assigned to this class and she decided to use CAS principally to aid students' conceptual understanding. She made sure that CAS was available at all times: in lectures, in the frequent laboratory sessions, and for all forms of assessment. Using an overhead projector, she carefully demonstrated relevant CAS techniques for new mathematical ideas and for further explanation of problems and issues arising during the lecture. Later, in the laboratory, the students worked through guided worksheets. Some of the activities involved learning essential new CAS techniques (Chapter 5), demonstrated on familiar mathematics whenever possible. Other worksheet activities involved exploration, investigation, and the solution of challenging problems.

As her experience of teaching with CAS grew, Claire noticed that her teaching style had become more student-centred. With CAS always available, she was able to change the focus of her interactions with students from the whole class to small groups and individuals. She also encouraged the students to discuss their work with each other and to make notes on the mathematics they were observing. Heid & al (1990) note that the roles for teachers teaching with technology include technical assistant, collaborator, facilitator of student learning, and catalyst. Claire believes that she had changed her role from lecturer to facilitator and catalyst, at least in part because of the support that the technology was providing for independent student work. She also believed that students' understanding was enhanced by their discussion with the teacher and other students, much of which seemed to include the computer as an extra authority contributing its own answers to the discussion (Pierce & Stacey 2001b). For example, a great deal of student-student discussion arose in Claire's classes because the students were still having difficulty with basic aspects of algebraic equivalence (Chapter 9 § 2). Students who expected an answer $\log(x^{-2})$ might comment *"but my computer says it is –2log(x): what does your computer say?"* Schneider (2000) made similar observations about the changing roles of teachers and students using CAS in an Austrian project: "Whereas in the non-CAS supported classroom many students actively participated in only half of the lessons this increased to three quarters of the lessons in the CAS-supported classroom". The literature suggests that these changes in teaching style and classroom environment are likely to impact favourably on students' achievement. For example, Keller & al (1999) report that student-centred teaching styles (independent of the type of symbolic calculator adopted) had a significant, positive effect on the numbers of students succeeding and the uniformity of success across ability levels.

Along with a change in teaching style, Claire moved from primarily using CAS functionally to primarily using it pedagogically. This different use of technology was reflected in the worksheets. Early worksheets were designed for students to take full advantage of the symbolic capabilities of DERIVE. For example, the first exercises on derivative were based on the function $f(x) = \sin(x)/x$. The students used DERIVE as a tool to extend their experience of mathematical functions. However, with experience, Claire began to include activities for three different purposes:

- some problems had the explicit purpose of introducing students to new CAS commands and syntax, and Claire found that problems that reviewed familiar material were most suitable for this;
- new procedures or concepts were introduced through activities (with a variety of exercises) involving one or more representations. These aimed to give students a sense of ownership through the guided discovery of basic patterns and concepts;
- students were still expected (permitted) to use CAS for more challenging questions and for computations that were more time consuming. For example, to teach the early rules for symbolic differentiation Claire set exercises that were now based on CAS use with simple power functions, for example, x^2, x^3, x^{-2}, x^{-5}, $x^{3/2}$, $x^4 + 2x^2 + 1$. Students used CAS to generate derivatives of these functions and then conjectured what the general rules may be (Figure 4-4).

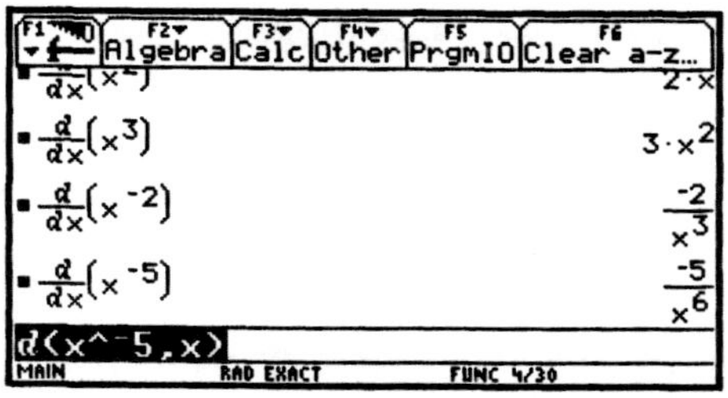

Figure 4-4. Looking for patterns to discover rules for differentiation

Other questions taken from textbooks often needed sophisticated techniques and required the students to delve more deeply into a topic.

After this guided discovery phase, Claire formally lectured the students using a traditional, paper-and-pencil approach. Understanding was developed through a series of exercises requiring students to explore multiple representations: rate of change, tangents to curves, and the derivative function. Finally, after being introduced to both the mathematics and the required CAS commands, the students (with CAS available), successfully tackled realistic and complicated investigative exercises some of which required paper-and-pencil techniques beyond most of these students' current capabilities. In Claire's first experiences, students complained about the workload involved in learning how to use CAS as a modern tool for the workplace. However, as the interfaces of CAS machines have become easier to use, and since many students now enter her classes already possessing significant relevant technology skills (e.g., knowing how to use a graphics calculator), the time spent directly learning to use the tool is now much less.

Finally, Claire also changed the focus of examinations by including more questions that required students to interpret and explain their answers rather than merely recall facts and skills. For example, in the 1997 examination six out of twelve questions involved interpretation compared with only two out of twelve questions on the 1996 examination. After being taught mathematics with CAS, the students were expected to move beyond routine symbolic mathematical manipulations, or translating conventional mathematical words and symbols into CAS commands. They were expected to be able to explain their working, verbalise and justify their reasoning, and interpret their results. Ability in paper-and-pencil computation or symbolic manipulation was no longer regarded as an adequate measure of mastery of a mathematical topic.

In summary, over a period of time, Claire changed her teaching approach using CAS. Initially, she had privileged the symbolic representation using paper-and-pencil techniques and symbolic algebra more than the graphical and numerical representations. With experience, she redesigned her lessons to enable students to learn mathematics while using CAS. She gave the students more opportunities to construct meaning for themselves, principally through revised worksheets that required exploration of mathematical patterns and multiple representations with relatively simple functions. Using CAS to deal with more complicated functions came later. Claire also changed the focus of examination questions. In sum, as Claire's experience in teaching with CAS has matured, her classroom use of CAS has shifted from a primarily functional use to a primarily pedagogical use.

6. TEACHING IN A TIME OF TRANSITION

All three of our teachers are pioneers, working in a time of transition from old to new ways of doing and teaching mathematics. Teachers need to support students' learning of both the technology and the mathematics, which are simultaneously changing. In this section we draw together insights from what the three teachers have done and the issues they have faced.

6.1 Teaching with new technology

Learning to use such a complicated machine as a symbolic calculator cannot be left to the student alone. All of the teachers developed their own styles of helping students and of managing the class using technology.

Andre showed considerable growth in his skills of teaching with technology during the project. His positive experience of the convenience of using the overhead projector and his admiration of the features of the symbolic calculator (including its exact answers, large screen, and clear menu structure) gave him confidence to try other technologies. His systematic, procedural approach to mathematics was evident in his use of flowcharts and notes about calculator procedures. He taught his students mathematical procedures and symbolic calculator procedures simultaneously, emphasizing both.

Benoit, in contrast, began the project already accomplished in teaching with graphics calculators. He too taught his students mathematical procedures and symbolic calculator procedures simultaneously, but emphasized the latter much less. It was not as important to Benoit that the students could use their symbolic calculators efficiently for symbolic manipulation. In our testing, we saw that his students under-utilised the machines and often made paper-and-pencil manipulation errors that symbolic calculator use would have avoided, although they demonstrated more conceptual understanding than Andre's class (Kendal & Stacey 1999, Kendal 2002). We suspected that Benoit's method of teaching technology use through class discussion and visualisation, with only minimal use of technological visual aids, worked only because of his exceptional classroom management skills.

Over several years, Claire has developed a teaching style for teaching with technology that has led students to adopt positive learning strategies and co-operation in the computer laboratories. This is not an isolated instance. Reporting on another Australian study, Forster (1997) commented on the development of a classroom environment where students interacted effectively with the teacher and each other. She concluded that, "technology-based learning was well suited to a student-centred exploratory approach and seemed most effective when the students worked collaboratively with each other". It remains to be seen whether this situation will remain the same for Claire now that her students are using hand-held symbolic calculators where they cannot easily see each other's work, rather than computers with screens visible to a group. She still uses the large screen for the visualisation and demonstrations to the whole group, so that full class discussion is unaffected, but the easy sharing amongst students may not survive the move from computer to calculator.

In summary, the different strategies that the teachers employed suggest that there will be a variety of successful solutions to the problem of teaching both mathematics and technology use. Trouche (Chapter 8) describes an alternative classroom structure that has been trialled in French classrooms and he proposes a framework that considers the role of classroom structure in technological environments.

6.2 Using time for mathematics or for technology?

Oldknow & Taylor (2000) are among the authors who believe that using CAS in school mathematics will save time and that "The teacher is able to use the time gained to extend the pupils' mathematical understanding". Is this the experience of our three teachers?

All three of the teachers have found that teaching CAS procedures at the same time as teaching the new mathematics content was possible. Benoit focused on assisting students to understand the new mathematics; he used the calculators alongside the students in a natural, matter of fact way without using any additional classroom equipment (overhead projector, for example) or new teaching strategies. He commented that, on occasions, using CAS saved time such as when data gathering for lessons based on an investigation of patterns. This additional time was absorbed into classroom discussion.

In earlier courses, Claire had allocated time specifically to learning to use CAS, but this became unnecessary as the CAS interface became easier, students came with more skills, and the focus of the university curriculum changed. The students spent the time saved on exploratory projects to enhance their understanding of the mathematics. Andre was the teacher who most freely permitted CAS techniques to be used by students; therefore he was in a position to gain most time by reducing practice of paper-and-pencil skills. However, he reallocated this time to teaching calculator procedures.

It is significant that Andre taught CAS procedures in a way that did not integrate the technology and the mathematics: he spoke and wrote notes about button sequences on the machine: *"press F4, then F6"*, and so forth. In contrast both Claire and Benoit spoke about the mathematical procedures using proper mathematical vocabulary: *Differentiate*, then *Solve*, and so forth. Claire believes that doing this helps her students acquire an overview of the mathematical method and not just routines that are specific to particular hardware or software. More importantly, the focus of the teaching is always on learning the mathematics with the CAS used as a means to this end.

The amount of time that is gained by use of CAS technology is therefore variable and is influenced by a range of factors. Stacey & al (2000a) observe that the best way in which to reallocate time needs to be seriously considered as part of the curriculum response to CAS use. It is apparent that after the transition period, time should become available for teachers to reallocate, but whether this is allocated to additional topics, to increase understanding, to develop better capabilities for formulating real problems in mathematical terms, or to some other goal is an important future choice.

6.3 Making full use of the symbolic facility

As we noted above, Benoit used the symbolic facility of CAS in a constrained way; he embraced its use only for generating data for students to guess patterns and rules. We explained above that, for several reasons, he was very cautious, especially in the second trial. On reflection, however, we see that the lessons we developed did not necessarily need symbolic algebra for solving problems. Within a couple of lessons of being introduced to a topic nearly all of the students could solve the problems using paper-and-pencil manipulation. There was some re-ordering of material, so that students could observe the numerical and graphical properties of differentiation before the formal procedures were taught, but beyond this, questions were quickly answered from within the expected student paper-and-pencil repertoire. The lack of real need to use CAS may have contributed to Benoit's decision to reduce the emphasis on using CAS for symbolic algebra. Whereas Claire's classes included some problems where the symbolic facility of CAS was needed, the CAS-Calculus project lessons did not involve symbolic skills beyond the expected paper-and-pencil level.

This lack of need for assistance in symbolic manipulation in the research project is in direct contrast to the need for machine graphing: numerous problems were used where tracing graphs on the machine was a precise aid. Graphing is conceptually a relatively simple procedure that is very tedious to carry out in practice without technology. Incorporating graphics calculators into teaching therefore has obvious benefits and can make problems easier and faster for students. In our schools, we are now seeing a rise in the number of students who solve problems graphically. For example, Charles (the third secondary school teacher in the first CAS-Calculus teaching trial) stressed a graphical approach to a wide range of problems. In our testing, his students used a high proportion of graphical methods: they were relatively better at solving problems than students who mainly used an algebraic approach (Kendal & Stacey 1999). It seems likely that problems that require the symbolic algebra facility of CAS will be perceived as more complicated or sophisticated than those in our standard curriculum. This issue (discussed in more detail in the next section) is receiving ongoing attention as we develop experience with the new subject that permits CAS in its external examinations (Stacey & al 2000b, Flynn 2001, 2003).

6.4 Finding the place of paper-and-pencil skills in a CAS curriculum

All three teachers faced decisions about which skills were essential for students to master using paper-and-pencil. This issue was the least problematic for Andre, who accepted the ability to carry out a routine procedure (such as differentiating) on the symbolic calculator as of equivalent value to the ability to carry it out with paper-and-pencil. Andre's procedural view of mathematics led him to accept that there are alternative procedures. In contrast, Benoit wanted 'understanding' and felt that implementing rules with paper-and-pencil (e.g., differentiating $x^5 + 2x^3$ and $(x^3 + 5)(3x^2-x)$) contributed to this to such an extent that it was irreplaceable. As we noted above, this was also fuelled by concern that his students needed to learn paper-and-pencil procedures for future, external, state-wide examinations.

Claire, whose students do not do external examinations, was freer to consider the mathematical worth of paper-and-pencil skills but still felt external pressures. Since some of her students would become teachers, she adopted a middle ground on teaching paper-and-pencil skills. Claire's students also tended to support this (Pierce & Stacey 2001b). Monaghan (1997) suggests that the more important algorithms are those that play an important part in students' cognitive development or those that contain a principle that is important for later development, but it is not an easy task to identify these. Claire encouraged students to perform routine skills exercises by paper-and-pencil when she believed they contributed to the students' understanding of the mathematics. However, until the external curriculum environment changes, teachers and students will live in an ambiguous situation about paper-and-pencil skills.

6.5 Changes required in schools

The experiences of Claire, Andre, and Benoit demonstrate that the introduction of CAS will present many challenges to educators. Some teachers will find it very difficult to make the changes to their teaching practices that will enable students to use CAS in a non-trivial fashion. Thomas & al (1995) report on a New Zealand study involving teachers using computers in their classrooms during one school year. They point out that:

> Putting a computer in the mathematics classroom is unlikely to result in changes in learning or teaching unless the personal philosophy of classroom practice held by each teacher undergoes a major transformation.

Andre and Benoit were volunteers who willingly explored new ideas, but they demonstrate how technology is adapted to individual teaching styles, rather than changing them. Teachers in other research (e.g., Simmt 1997, Tharp & al 1997) showed a related behaviour. Although they made changes to their teaching styles in their first attempt at teaching with graphical calculator technology, they showed a tendency to return, in the second attempt, to their former methods. For the CAS-Calculus project, the teachers' ability to change their teaching styles was exacerbated by the fact that the CAS-Calculus project was conducted as a unique trial, in isolation. The use of CAS was not institutionalised within the particular school and this made it difficult for teachers to endorse its use fully (Kendal & Stacey 2002). In contrast, the university teachers were able legitimately to introduce CAS into their curriculum and gradually change their teaching program because they had control over the accreditation and assessment of the courses. The teachers, including Claire, had the opportunity to make significant changes to their pedagogy gradually.

For the wider school system, substantial professional development will be needed. Research on the introduction of technology in schools points to the importance of suitable professional development programs that explain the rationale for change, support teachers' learning how to use the new technology, and show teachers how to teach with technology effectively. Kutzler (2003) also believes that if teachers learn about good teaching practices with technology, through high quality professional development, they may be motivated to change their previous inappropriate pedagogy and become successful in teaching good mathematics with the assistance of CAS. These issues are further discussed in the conclusions and perspectives chapter of this volume.

Some students also experience difficulties adapting to technology and changing their style of learning. In the CAS-Calculus project, the students were familiar with paper-and-pencil methods and using a graphics calculator. Some students were unable or unwilling to change during the CAS-Calculus project. Voigt (1994) observes that hidden and stable regularities of classroom life persist when teachers want to introduce change.

Teachers have the responsibility to select appropriate curriculum materials that help students learn and succeed in the new CAS environment (Zbiek 2003). Thus, changes will need to be made to existing curriculum but there is a range of possibilities about exactly what changes should be implemented. Stacey & al (2000a) canvas options for goals of using CAS in senior mathematics subjects. They contend that CAS can be introduced into a school curriculum for a variety of purposes: in order to align school mathematics with the use of technology in the modern world, to make students better users of mathematics, to help students achieve deeper learning, or to free up curriculum time so that new topics can be studied. However, they comment that not all these goals can be achieved at once and school systems must choose some at the expense of others.

Changes will also need to be made to student worksheets and textbooks to accommodate changes in curriculum and teaching practices. Claire showed that it is possible to modify pre-CAS worksheets to make them more suitable for use in a CAS supported classroom. The book by Berry & al (1997) shows a range of approaches to these tasks.

Changes must also be made to assessment if CAS is to be successfully implemented into school curricula. Assessment drives students' learning and much of teachers' teaching. All participants come to value and concentrate on what will be assessed. Students who cannot use CAS in their examinations are not likely to make a substantial investment of time and effort to learn how to use it well. If CAS is to be taken seriously as a tool for learning mathematics then students need to have access to it for assessment tasks. In turn, what is assessed has to reflect what has been taught and learnt. Paper-and-pencil calculation has been used as a test of understanding even though all the evidence points to it being a limited indicator. Now the challenge is to design a broader range of items that focus on widening the bandwidth of mathematical knowledge assessed and better test symbolic reasoning (for example, Flynn 2001, 2003). Flynn (2001) recommends that:

> CAS-permitted examinations be on balance accessible to all CAS users and contain examination questions that are appropriately worded and structured and take into account

the capabilities of CAS. Marking schemes need to be constructed along principles consistent with the way students document their solutions in such examinations. CAS-permitted examinations should allow some newer-style questions to be asked and widen the range of mathematical skills assessed currently.

As described above, Claire's revised assessment included more questions involving interpretation and explanation rather than routine symbolic mathematical manipulations. This is a key observation.

Finally, there is substantial evidence accumulating that students do not make enough use of technology when it is permitted in examinations. For example, studies in quite different contexts (e.g., Forster & Mueller 2000) show that students made insufficient use of the graphics calculators permitted in their examinations. This was also observed in the CAS-Calculus project and in examinations at the university. Some students reported that they felt that they were cheating when they used CAS for questions they had not completely mastered. For example, Claire gave her students an assessment task that required them to examine the function g(x) = sec(x) and the associated family of translated and dilated functions. The function sec(x)[5] is available on DERIVE, but this terminology was not known to Claire's students. They could however enter g(x) = sec(x) into DERIVE and use standard techniques for translations and dilations. Although Claire expected them to do this, some students did not attempt the question while others who were observed to graph the function with DERIVE in the examination did not record their answer, subsequently stating that they did not trust the computer for a function they did not recognise. Teachers need to ensure students realise the potential of the permitted technology.

Soon, every student in first-world countries may have access to CAS for learning mathematics. This chapter has shown that using CAS in the classroom provides many opportunities for improving student learning. However, taking advantage of powerful software requires changes to many aspects of the classroom and many players share the responsibility of making sure that CAS use benefits students:

- education researchers have the responsibility to identify the key variables and disseminate findings on effective practice to educators. To advocate change there should be substantial body of evidence that supports the innovation and the ways that students learn with technology need to be understood;
- technology designers can make a great difference by providing interfaces that are suitable to support teaching and learning;
- schools will need to support careful changes in curriculum design and assessment: they will need to encourage teachers to undertake training to teach with CAS;

- teachers will have the opportunity to adopt more student-centred teaching practices, and have the responsibility to provide students with suitable tasks that help them to construct knowledge for themselves.

7. CONCLUSIONS

With CAS, students have the opportunity to fulfil their mathematical potential with less computational burden. Using suitable teaching materials, competent teachers can focus student attention on the meaning of the mathematics under consideration. CAS can assist teachers to enhance students' opportunity to acquire insightful problem solving skills, develop deep conceptual understanding, develop higher levels of thinking, and gain an understanding of how to validate and interpret solutions. CAS technology can prove to be a powerful mathematical partner.

The experiences of our pioneering teachers show some of the first steps along the road towards this ideal situation. The classrooms of Andre and Benoit illustrate how current differences between teachers will not disappear and may even be exaggerated by intelligent tools. The technology supports learning and teaching of many different styles, including both teaching emphasizing routine procedures and teaching emphasizing understanding. On the other hand, Claire's experiences demonstrate an evolution that has been driven by a thoughtful reaction to gradually emerging possibilities. Whereas graphics calculators, for many teachers, slotted easily into the curriculum and enhanced their teaching with little threat, CAS demands a more thorough response. Neither Andre nor Benoit, working in a somewhat artificial environment, were able to progress far along this track in the short time of the research project. Their first reactions were interestingly different on three points: in the ways they allowed it to change the curriculum (especially how they came to regard paper-and-pencil skills), what they valued when they taught with it, and how they managed their classrooms. The task for educators is now to move ahead simultaneously on curriculum and assessment, teaching styles and classroom organization, and more generally to the problems posed for teaching mathematics in the information age.

ACKNOWLEDGEMENTS

We gratefully acknowledge the cooperation of the teachers and students who participated in the research studies reported in this chapter.

Professor Kaye Stacey < k.stacey@unimelb.edu.au > is Foundation Professor of Mathematics Education at the University of Melbourne, Australia. Her research interests centre around students' mathematical thinking and problem solving, mathematics curriculum issues, and the impact of new technology on curriculum, assessment and teaching. She has a keen interest in teacher education, teaching undergraduate and post-graduate students.

After many years as a secondary school mathematics teachers, Dr Margaret Kendal < kendal@bacchusmarsh.net.au > has recently completed her doctoral thesis at the University of Melbourne, Australia. Using detailed observations of classroom teaching with CAS, she has tracked how teachers' curriculum choices and teaching style influence students' learning with new technology.

Dr Robyn Pierce < r.pierce@ballarat.edu.au > is a teacher and researcher at the University of Ballarat, Australia (School of Information Technology and Mathematical Sciences). She teaches a variety of mathematics and statistics subjects, within which she has pioneered teaching with CAS and other new technology. Her recent research work is documenting the development of algebraic insight and helping students make effective use of CAS.

APPENDIX

Appendix 4-1
Benoit, Andre, Claire... and others
Recent studies about teachers using technology
(Jean-Baptiste Lagrange)

A review of literature on the use of technology in the teaching of mathematics (Lagrange & al 2003) shows that there was very little research on the teacher from 1994 to 1999. At that time, the main focus of interest was the instructional possibilities offered to the teachers through the use of technology. Chapter 4 adopts a different approach. It is a study of the constraints and opportunities introduced by the use of technology, from the point of view of the teacher and in light of their conceptions of teaching and of mathematics.

Zbiek (2001) comments that experienced teachers have reached a certain ease in their teaching practices, which is disturbed by the introduction of computer algebra; allowing, and demanding at the same time, new decisions, new actions, and a new understanding. Thus, the integration of computer algebra into teaching not only introduces new classroom variables but it also changes the options for existing variables, which can increase the differences between teachers. This approach to studying teaching may reveal, in a particularly clear way, the impact of computer algebra on teaching. Eventually, as Zbiek (ibid.) emphasizes, it may lead to the identification of particular strategies that make it easier for the teacher to teach with computer algebra.

Among the rare studies based on this approach, Lumb, Monaghan & Mulligan (2000) report the successes and the problems met by two teachers, Steve and Stephen, when they tried to intensively use the computer algebra software, DERIVE. It is obvious that the introduction of a new technology requires the teacher to devote considerable time to lesson preparation, but according to Steve and Stephen, compared to other teaching software, DERIVE is at the upper end of the 'effort' scale. Moreover, it takes longer to get a 'feel' for how to use DERIVE, and the authors attribute this to the extensive possibilities afforded by this software. In addition, Bottino & Furinghetti (1998), state that, even with experience, the teachers exploit only a small part of the software's potential.

Steve admits that many of his first ideas for exploiting DERIVE are actually not feasible. An analysis of the activities he deployed in class shows, among other changes, that he reduced the time he spent on what the authors call 'coaching' activity. 'Coaching' consists in engaging the students in thinking and reasoning about the mathematical situation under consideration, without telling them the key ideas. This reduction is paradoxical, since the introduction of computer algebra was supposed to allow more time to be spent directly on mathematical ideas. In Monaghan's study (2001) of thirteen teachers, he notes, that in the technology based lessons, the 'coaching' activity was reduced and replaced by another 'coaching', this one related to the technical aspects of the use of the software. He notes the idea that the teacher could play the role of a 'catalyst for self-directed student learning' seems not to work.

This paradoxical decrease of the 'coaching' is obviously related to the teachers' inability to 'feel' how to use the software for teaching and learning. Indeed, in order to elicit mathematical insights from students, the teacher needs to anticipate learning opportunities. Without technology, experienced teachers are able to improvise easily on their established strategies. In contrast, with technology and especially with computer algebra, it is not always immediately obvious to the teacher how to exploit the situation mathematically and his previous ease in teaching is called into question. It is then easier for him to play the technical assistant than to explain about mathematics.

Steve and Stephen put a lot of effort into preparations of worksheets for students, deviating from their usual practice of using a text book because they believed it was not possible to adapt much of the text book material for use with the technology. Neither did they use many of the materials produced by the research group in which they participated. As stated by the authors of the article (p.236): "... we think that teachers who plan to incorporate a significant use of computer algebra in their teaching are presented with a re-evaluation of the mathematics they were taught, and are familiar with".

Thus, trying to really integrate computer algebra is not easy for Steve and Stephen, as for Benoit, Andre, and Robyn. Moreover "it may be perceived as something that is neither rewarding nor desirable" (Lumb, Monaghan & Mulligan, ibid., p.239). Nevertheless their experience gives a fair idea of the work necessary for the successful integration of CAS into teaching. In chapter five, drawing on observations of more teachers (Schneider 1999), we interpret this work as the difficult elaboration of new *praxeologies* (Box 5-1). These include not only some different mathematics, but also the new decisions and the new actions mentioned by Zbiek (ibid.).

NOTES

1. The word 'calculus' refers to the first course in 'analysis' in the curriculum of high schools in many English speaking countries.
2. We acknowledge the work of Barry McCrae and Gary Asp, who were other researchers working on this project.
3. Claire is a pseudonym for one of the three authors.
4. Andre and Benoit's students were all girls.
5. Sec(x) is standard notation for 1/cos(x).

REFERENCES

Berry J., Monaghan J., Kronfellner M. & Kutzler B. (Eds.), (1997) *The state of computer algebra in mathematics education*. Lund, Sweden: Chartwell-Bratt.

Bottino R. & Furinghetti F. (1998) The computer in mathematics teaching: scenes from classroom, in D. Tinsley & D.C. Johnson (Eds.), *Information and communication technologies in school mathematics* (pp. 131-139). London: Chapman & Hall.

Flynn P. (2001) Examining mathematics with CAS: Issues and possibilities, in C. Vale, J. Horwood & J. Roumeliotis (Eds.), *A mathematical odyssey, Proceedings of the 38th annual conference of the MAV* (pp. 285-298). Melbourne: Mathematical Association of Victoria.

Flynn P. (2003) Adapting "Problems to prove" for CAS-permitted examinations, *The International Journal of Computer Algebra in Mathematics Education* 10(3), 103-121.

Forster P. (1997) Teaching through technology, *Australian Senior Mathematics Journal* 11(1), 54-61.

Forster P. & Mueller U. (2000) Diversity and difficulties with graphics calculator usage, in the 1999 calculus tertiary entrance examinations in Western Australia, *Australian Senior Mathematics Journal* 14(1), 4-15.

Heid M.K. (1988) Resequencing skills and concepts in applied calculus using the computer as a tool, *Journal for Research in Mathematics Education* 19(1), 3-25.

Heid M.K., Sheets C. & Matras M.A. (1990) Computer enhanced algebra: New roles and challenges for teachers and students, in T. Cooney (Ed.), *Teaching and learning mathematics in the 1990s, NCTM 1990 Year Book*. Reston, VA: NCTM, 194-204.

Keller B., Russell C. & Thompson H. (1999) A large-scale study clarifying the roles of the TI-92 and instructional format on student success in calculus, *The International Journal of Computer Algebra in Mathematics Education* 6(3), 191-207.

Kendal M. & Stacey K. (1999) Varieties of teacher privileging for teaching calculus with computer algebra systems, *International Journal of Computer Algebra in Mathematics Education* 6(4), 233-247.

Kendal M. & Stacey K. (2001) The impact of teacher privileging on learning differentiation with technology, *The International Journal of Computers for Mathematical Learning* 6, 143-163.

Kendal M. (2002) *Teaching and learning introductory differential calculus*, Unpublished doctoral thesis. Australia: The University of Melbourne. Available: (http://thesis.lib.unimelb.edu.au/).

Kendal M. & Stacey K. (2002) Teachers in transition: Moving towards CAS-supported classrooms, *Zentralblatt für Didacktik der Mathematik* 34(5), 196-203.

Kutzler B. (2003) CAS as pedagogical tools for teaching and learning mathematics, in J. Fey, A. Cuoco, C. Kieran, L. McMullin, & R.M. Zbiek (Eds.), *Computer algebra systems in secondary school mathematics* (pp. 53-71). Reston: National Council of Teachers of Mathematics.

Lagrange J.B., Artigue M., Laborde C. & Trouche L. (2003) Technology and Mathematics Education: a Multidimensional Study of the Evolution of Research and Innovation, in A.J. Bishop, M.A. Clements, C. Keitel, J. Kilpatrick & F.K.S. Leung (Eds), *Second International Handbook of Mathematics Education* (Vol. 1, pp. 239-271). Kluwer Academic Publishers.

Lumb S, Monaghan J. & Mulligan S. (2000) Issues arising when teachers make extensive use of computer algebra, *The International Journal of Computer Algebra in Mathematics Education* (7)4, 223-240.

McCrae B., Asp G. & Kendal M. (1999) Learning calculus with supercalculators, in J. Truran & K. Truran (Eds.), *Making the difference, Proceedings of the 22nd annual conference of The Mathematical Educational Research Group of Australasia* (pp. 359-364). Adelaide: MERGA.

Monaghan J. (1997) Teaching and learning in a computer algebra environment: Some issues relevant to sixth-form teachers in the 1990's, *The International Journal of Computer Algebra in Mathematics Education* 4(3), 207-220.

Monaghan J. (2001) Teachers' classroom interactions in ICT-based mathematics lessons, in M. Van den Heuvel-Panhuizen (Ed.), *Proceedings of PME 25* (Vol. 1, pp. 383-390). Utrecht: Freudenthal Institut.

National Council for Educational Technology (1994) *The A-level curriculum of the future - The impact of computer algebra systems.* Coventry: NCET.

Oldknow A. & Taylor R. (2000) *Teaching mathematics with ICT.* London and New York: Continuum.

Palmiter J.R. (1991) Effects of computer algebra systems on concept and skill acquisition in calculus, *Journal for Research in Mathematics Education* 22(2), 151-156.

Pierce R. & Stacey K. (2001a) Reflections on the changing pedagogical use of computer algebra systems: Assistance for doing or learning mathematics? *Journal for Computers in Mathematics and Science Teaching* 20(2), 141-159.

Pierce R. & Stacey K. (2001b) Observations on students' responses to learning in a computer environment, *Mathematics Education Research Journal* 13(1), 28-46.

Repo S. (1994) Understanding and reflective abstraction: learning the concept of derivative in a computer environment, *International DERIVE Journal* 1(1), 97-113.

Schneider E. (1999) La TI-92 dans l'enseignement des mathématiques, des enseignant(e)s découvrent la didactique des mathématiques, in D. Guin (Ed.), *Calculatrices symboliques et géométriques dans l'enseignement des mathématiques, Actes du colloque francophone européen* (pp. 49-60). Montpellier: IREM, Université Montpellier II.

Schneider E. (2000) Teacher experiences with the use of CAS in a mathematics classroom, *International Journal of Computer Algebra in Mathematics Education* 7(2), 119-141.

Simmt E. (1997) Graphing calculators in high school mathematics, *International Journal of Computers in Mathematics and Science Teaching* 16(2/3), 269-289.

Stacey K., Asp G. & McCrae B. (2000a) Goals for a CAS-active senior mathematics curriculum, in M. Thomas (Ed.), *TIME 2000, Conference proceedings at an International Conference on Technology in Mathematics Education, 11th-14th December* (pp. 244-252). Auckland: The University of Auckland.

Stacey K., McCrae B., Chick H., Asp G. & Leigh-Lancaster D. (2000b) Research-led policy change for technologically-active senior mathematics assessment, in J. Bana & A. Chapman (Eds.), *Mathematics Education Beyond 2000, Proceedings of the 23rd annual conference of the Mathematics Education Research Group of Australasia* (pp. 572-579). Fremantle: MERGA.

Tharp M.L., Fitzsimmons, J.A. & Brown Ayers, R.L. (1997) Negotiating a technological shift: Teacher perception of the implementation of graphing calculators, *Journal of Computers in Mathematics and Science Teaching* 16(4), 551-575.

Thomas M., Tyrrell J. & Bullock J. (1995) Using computers in the mathematics classroom: The role of the teacher, *Mathematics Education Research Journal* 8(1), 38-57.

Victorian Curriculum and Assessment Authority (Accessed 3 Sep 2003). Available: (http://www.vcaa.vic.edu.au/vce/studies/MATHS/).

Voigt J. (1994) Negotiation of mathematical meaning and learning mathematics, *Educational Studies in Mathematics* 26, 275-298.

Zbiek R.M. (2001) Influences on Mathematics Teachers' Transitional Journeys in Teaching with CAS, *CAME Meeting*. Utrecht: (http://ltsn.mathstore.ac.uk/came).

Zbiek R. M. (2003) Computer algebra systems in secondary schools mathematics education, in J. Fey, A. Cuoco, C. Kieran, L. McMullin & R.M. Zbiek (Eds.), *Computer algebra systems in secondary school mathematics* (pp. 197-216). Reston: National Council of Teachers of Mathematics.

Chapter 5

USING SYMBOLIC CALCULATORS TO STUDY MATHEMATICS

The case of tasks and techniques

Jean-Baptiste Lagrange
IUFM de Reims & DIDIREM, Université Paris VII, France
jb.lagrange@reims.iufm.fr

Abstract: This chapter will consider in more depth the possible contribution of technology -- especially CAS -- to the study of mathematical domains. Using a theoretical approach to treat examples of classroom activities, we will show how a didactical reflection can help to understand this contribution. A variety of new techniques will be presented and related to paper-and-pencil techniques. Examining the pragmatic and epistemic value of both types of technique will help to make sense of classroom situations. It will also help to clarify the situation of teachers wanting to integrate new tools. Consideration of other approaches will show that educators emphasize the use of computer algebra to promote 'conceptual' mathematics. Nevertheless, they cannot ignore instrumented techniques when considering the real potentialities of new tools and the conditions for their integration.

Key words: Conceptualization, Tasks, Techniques.

1. INTRODUCTION

Using new tools, students can now easily perform numerical and symbolic calculation that would be very painstaking by hand. As we saw in Chapter 3, transposing experimental approaches from mathematical sciences into teaching seems to be a realistic and stimulating prospect, but the question of the contribution that experimental approaches inspired by mathematical sciences might bring to students' conceptualizations remains open. We concluded that this question would require the addressing as a whole of the study of a mathematical domain and the way in which it is changed by new approaches that tools make possible.

Box 5-1.

The anthropological approach

(Artigue 2002)

The anthropological approach (Chevallard 1999) shares with socio-cultural approaches in the educational field (Sierpinska & Lerman 1996) a vision of mathematics as the product of a human activity. Mathematical productions and thinking modes are thus seen as dependent on the social and cultural contexts where they develop. As a consequence, mathematical objects are not absolute objects, but are entities, which arise from the practices of given institutions. The word *institution* has to be understood in this theory in a very broad sense: the family is an institution for instance. Any social or cultural practice takes place within an institution. Didactic institutions are those devoted to the intentional apprenticeship of specific contents of knowledge. As regards the objects of knowledge which it takes in hand, any didactic institution develops specific practices, and this results in specific norms and visions as regards the meaning of knowing or understanding such and such an object. Thus to analyze the life of a mathematical object in an institution, to understand the meaning in the institution of 'knowing/understanding this object', one needs to identify and analyze the practices which bring it into play.

These practices, or *praxeologies*, as they are called in Chevallard's approach, are described by four components: a type *of task* in which the object is embedded; the *techniques* used to solve this type of task; the *technology*, that is to say the discourse which is used in order to both explain and justify these techniques; and the *theory* which provides a structural basis for the technological discourse itself and can be seen as a technology of the technology. Since we have already assigned a meaning to the word technology in this book, so as to avoid misunderstanding, in the following we combine Chevallard's *technological* and *theoretical* components into a single *theoretical* component. The word *theoretical* has thus to be given a wider interpretation than is usual in the anthropological approach. Note also that the term *technique* has to be given a wider meaning than is usual in educational discourse. A *technique* is a manner of solving a task and, as soon as one goes beyond the body of routine tasks for a given institution, each technique is a complex assembly of reasoning and routine work. We would like to stress that techniques are most often perceived and evaluated in terms of *pragmatic value*, that is to say, by focusing on their productive potential (efficiency, cost, field of validity). But they have also an *epistemic value*, as they contribute to the understanding of the objects they involve, and thus techniques are a source of questions about mathematical knowledge.

For obvious reasons of efficiency, the advance of knowledge in any institution requires the routinization of some *techniques*. This routinization is accompanied by a weakening of the associated theoretical discourse and by a 'naturalization' or 'internalization' of associated knowledge which tends to become transparent, to be considered as 'natural'. A *technique* which has become routine in an institution tends thus to become 'de-mathematized' for the members of that institution. It is important to be aware of this naturalization process, because through this process *techniques* lose their mathematical 'nobility' and become simple acts. Thus, in mathematical work, what is finally considered as mathematical is reduced to being the tip of the iceberg of actual mathematical activity, and this dramatic reduction strongly influences our vision of mathematics and mathematics learning and the values attached to these.

The anthropological approach opens up a complex world whose 'economy' obeys subtle laws that play an essential role in the actual production of mathematical knowledge as well as in the learning of mathematics. A traditional constructivist approach does not help us to perceive this complexity, much less to study it. Nevertheless, this study is essential because, as pointed out by Lagrange (2000), it is through practices where technical work plays a decisive role that one constructs the mathematical objects and the connections between these that are part of conceptual understanding.

First we have to define what we mean by the *study* of a domain. This notion comes from Chevallard (1999): to study a domain is to do mathematical work on this domain for educational purposes. In education as in professional research, working on a mathematical domain is trying to solve a set of problems through using and creating concepts. An important issue is how concepts are produced. As far as we can say in general, the work of a researcher is to structure a domain so as to make good questions appear. Good questions are not just problems, but specific questions that can be addressed in a mathematically appealing way. Concepts appear when structure becomes visible. Their formulation is the product of further structuring work.

In a teaching and learning context, we consider that working in a mathematical domain is done at three structural levels. The first level is that of *tasks*. Here, tasks are taken not just to be individual problems but rather as more general structures for problems. For instance, consider the domain of real functions. Problems can be expressed enactively from 'real life' situations. A task like "find the intervals of growth of a given function" constitutes a common reference for some problems but not for others; likewise another relevant reference task is "find the zeros..."

Techniques are the second structural level. Technique has to be taken in the general sense of "a way of doing tasks". Techniques help to distinguish and reorganize tasks. For instance different techniques exist for the task "find the intervals of growth of a given function" depending on what is known about the function. If the function is differentiable the task can then be related to the task "find the zeros" of another function. In other cases, a search based on a more direct algebraic treatment can be more effective.

The third level is that of *theories*. While the first two levels are related to action, this level is related to assertion. At this level, the consistency and effectiveness of techniques are discussed. Mathematical properties, concepts and a specific language appear.

In some respects, this three level structure defining the study of a domain has to be taken as a postulate. On the one hand, doing mathematics in a domain necessarily involves structuring problems in terms of concepts. On the other hand, why choose to focus on tasks and techniques as intermediate structures? The reason is that, as we saw in Chapter 3, the potentialities of technologies -- and especially of CAS -- are expressed in terms of the expanded possibilities of action in solving problems. Thus, if access to concepts is seen as depending on the possibilities of investigating problems, technology should automatically enhance this access. We have stressed that things are not so simple. Observing several windows does not necessarily stimulate multi-representation thinking. Easily obtained symbolic results do not automatically provoke real inductive activity. The postulate we make in this chapter is that taking the above structural levels of tasks and techniques into consideration can account for these difficulties and help to think better about the support that technology can bring. Such a perspective has its origin in our surprise that consideration of tasks and techniques is often omitted in technological innovations, whereas they have an important place in 'real life' teaching and learning. Like all postulates, it will be justified inductively by its productivity.

2. THE IMPACT OF TECHNOLOGY ON STUDY

Our interest in techniques led us towards the anthropological approach developed by Chevallard (1999). The main elements are presented in Box 5-1. From these, the impact of computers on teaching and learning can be thought of at the level of techniques: traditional paper-and-pencil techniques are challenged by 'push button' techniques while, as we shall see, the use of technology requires *new techniques* dependent on the tool. The *pragmatic* and *epistemic* value of traditional techniques (Box 5-1) have to be reconsidered and new techniques have to be examined for a possible epistemic contribution.

To facilitate understanding, let us take the example of a small *praxeology* (Box 5-1) and examine the impact of a tool on techniques. Chevallard (ibid., p.243) considers the domain of expressions like $\frac{a+b\sqrt{2}}{c+d\sqrt{2}}$, *a*, b, c, and d integers. The study of this domain can be seen as a praxeology whose central task is the reduction of such expressions into $\alpha+\beta\sqrt{2}$, α and β rationals. The technique to accomplish this task is to multiply numerator and denominator by a suitable expression to obtain an integer denominator. This (tedious) technique provides for a *pragmatic* canonical writing of expressions from the domain, helping for instance to recognize that quotients like $\frac{\sqrt{2}-2}{3\sqrt{2}-4}$ and $\frac{\sqrt{2}-1}{2\sqrt{2}-3}$ are equal. Since performing the technique includes several elementary actions, each action implying an algebraic analysis of the expression -- especially before it has been routinized --, it can play an *epistemic* role (Box 5-1) towards developing knowledge of algebraic properties of quotients and radicals. At a theoretical level, it is a basis for the field structure of the algebraic extension $\mathbb{Q}[\sqrt{2}]$ and provides an algorithm to transform expressions into the canonical form.

In contrast a symbolic calculator accomplishes the reduction in just one operation. Using a paper-and-pencil technique, a human being will necessarily limit the number of expressions s/he will try, and focus on the underlying *algebraic property of radicals*. The use of a symbolic calculator for this task makes it possible to do more examples and orients the activity towards *pattern discovery* -- for instance recognizing that every quotient can be expressed as the sum of a rational and a rational multiplied by $\sqrt{2}$, that the expression is rational whenever ad = bc...-- and *generalization* building a praxeology for $\mathbb{Q}[\sqrt{k}]$ or $\mathbb{Q}[\sqrt[3]{2}]$. Clearly, the paper-and-pencil technique is linked to knowledge of elementary algebraic properties while the symbolic calculator technique opens up stable structures more directly, while hiding properties explaining the stability.

3. 'PUSH BUTTON' TECHNIQUES AND THEIR EFFECT ON CONCEPTUALISATION

CAS was created to ease the simplification of most common symbolic expressions. Corresponding tasks can be performed without great reflection. Thus, for students, 'push button' techniques tend to predominate over ordinary more painstaking techniques. For instance, after overcoming syntactical difficulties, an 11th grader will get limits even as simple as $\lim_{x \to +\infty} \frac{1}{x+1}$ by using the symbolic calculator *Limit* command with less pain than by reasoning. While by reasoning, s/he would have to think of a graphic asymptotical representation of the function $x \mapsto \frac{1}{x+1}$ or of a bounding by $x \mapsto \frac{1}{x}$, with a calculator s/he only has to enter *Limit*(1/(x+1),x,∞). Students adopt 'push button' techniques like this because of their simplicity and efficiency and there is a chance that they will link a concept too closely with the corresponding technique (Monaghan & al 1994).

This is an example that, with new tools, painstaking paper-and-pencil techniques retain little pragmatic value because they are challenged by 'push button' techniques. Routinization is no longer a necessity and thus their epistemic value could become more visible. However, paper-and-pencil techniques tend to become obsolete because of the ease of using CAS commands. This obsolescence is a problem because traditional techniques can no longer play their role in conceptualization and 'push button' techniques cannot take over this role directly.

Mathematics education has thus to reconsider the study of a domain, taking the obsolescence of traditional techniques into account, and to conceive new techniques as components of new praxeologies for this domain. Thinking of new techniques linked to the use of computer tools and of their possible epistemic value is not easy because mathematical culture is implicitly linked with paper-and-pencil techniques and is not accustomed to the idea that other tools can support conceptualization. However, this is indeed possible, as the next section will show.

4. A CAS TECHNIQUE AND ITS EPISTEMIC VALUE

The situation described in Box 5-2 illustrates how a technique linked to the use of CAS can be more than just 'pushing a button' and how it contributes to students' problem solving activity *and* mathematical conceptualization. Two teachers trialed three versions of the same situation.

The first version was deceptive as students worked with paper-and-pencil and could not go far in the value of n and thus in their conjectures. In the second version the teachers tried to use the CAS DERIVE to liberate students *"from the technical aspects of paper-and-pencil computing"* while encouraging them *"to keep sight of the main goal"*. While students actually engaged in experimental activity and learned about algebraic facts (degree of factors...), the situation did not produce real conjectures in spite of the teachers' expectation that students could find motivating conjectures by observing CAS factorizations with enough detachment so that they would not miss general factorizations.

Box 5-2.

The factorization of $x^n - 1$

(Mounier & Aldon 1996)

11th grade students (scientific stream) were asked to conjecture and prove 'general' -- true for every n -- factorizations of these polynomials by observing examples for some values of n. Three successive versions of this situation were developed:

Version 1: solving this problem with paper-and-pencil in a single session
Students found easily that x – 1 is a factor, then they used polynomial division to obtain a factorization for two or three values of n and were able to generalize into $(x-1)(x^{n-1}+x^{n-2}+...+x+1)$. Polynomial division was a tedious manipulation and after that, students did not look for other factorizations.

Version 2: with CAS in a single session
Students had to observe a set of outputs from DERIVE's *Factor Rational* command. A difficulty is that this command gives a most factored form while the general factorizations expected by the teachers are not complete for every n. For instance, the two-factor decomposition above is obtained only for prime values of n.
This is how students typically behaved. They used the *Factor Rational* command for n=2 and 3 and conjectured the above two-factor general factorization. Factoring for n=4 they thought that even n are not regular and trying n = 5, 6, 7 provided confirmation. At this point, they conjectured that there were two separate general factorizations, the above for odds and a three-factor one for evens. Wanting a confirmation for n = 9, they got a three-factor decomposition $(x-1)(x^2+x+1)(x^6+x^3+1)$. The conjecture was then rejected and students tried a variety of new conjectures but without success because they always found anomalous values of n. They did not go farther because a theory of cyclotomic polynomials that would explain DERIVE's factorizations was beyond their reach.

Version 3: a long-term problem
Students had access to Derive on laptops in the classroom and for personal work. The teachers assigned the factorization of x^n – 1 as 'a long term problem'. A first session introduced the problem and students were initiated to techniques for manipulating factors in DERIVE. Then students practiced at home and found conjectures and proofs that they reported in classroom discussions. For instance they recognized that the above factorization is true for all n and they proved factorizations like $x^{2^n}-1=(x-1)(x+1)(x^{2^2}+1)(x^{2^{n-1}}+1)$.

Students could not actually distinguish between DERIVE's and 'general' factorizations because they could not grasp the following idea: for a given n, several factorizations may exist, but only some of them are 'general' or true for a large set of values of n.

This situation is remarkable because students will have learned much about algebra if they can say, *"Yes, for given polynomials DERIVE gave us factored forms that are not the general factorization, but the general factorization is still valid because we can find it by collecting and expanding parts in the factored forms".* An important point to emphasize is that students who are able to make this statement know what it means to collect and expand parts of an expression. Mathematicians may think that this *technique* is obvious because they recognize complete and incomplete factorizations and understand that CAS provides the means to pass from one form to another.

To collect and develop several factors in a factorization is not so easy a manipulation for students: one must understand software-specific copy and paste functionalities and link them with an understanding of the structure of a factorized expression. In the situation reported above, this 'DERIVE technique' was missing, and this arose both from an insufficient knowledge of DERIVE *and* from a lack of understanding of the concept of factorization. Classroom elaboration of this technique is a condition for enabling experimental activity on the part of students *and* for giving this activity a mathematical dimension.

This observation illustrates what we said in Chapter 3. Experimenting 'like professional mathematicians' with the help of new tools is not so easily transposed into education. When not enough emphasis is put on techniques specific to the tool, a potentially rich situation may fail to bring students to conceptualizations. We observed that teachers using CAS were often reluctant to give time to these techniques. Since working regularly in a computer room presented difficulties, most of the teachers taught only a few sessions with DERIVE. In this context, they saw little *pragmatic* (Box 5-1) use for DERIVE techniques and tried -- often unsuccessfully -- to focus on conceptual issues. In contrast, experienced teachers and researchers successfully integrated DERIVE techniques into the classroom.

In a third version of the situation, introducing students to techniques for manipulating factors in DERIVE and making it possible for them to practice at home was more productive as the final report (Mounier & Aldon 1996, p.59) illustrates: for instance, students found and proved a non-trivial factorization where n is a power of 2. This proof by induction uses the expansion of a part of a factorization, a finding obviously linked to the technique. According to the authors, students did learn DERIVE as a new tool *and* changed their image of the concept of factorization. These teachers recognized the need to build techniques for using DERIVE and the epistemic role of these techniques in understanding algebra.

A provisional conclusion is that there is a great variety of new techniques, including 'push button' techniques and techniques needed to manage expressions. Obviously, in a paper-and-pencil context, one cannot bypass such techniques because of their pragmatic utility and one can easily overlook the epistemic value of such techniques. That is why recognizing new techniques and their epistemic value is not obvious for mathematics educators. In our analysis, instead of trying to reduce their importance or to bypass them, teaching has to considerer their pragmatic and epistemic value and their evolution during the mathematical work in a domain, in order to understand how the use of technology can support conceptualization.

In the next sections we will have a closer look at the variety of new techniques, emphasizing possible specificities, and we will consider how this approach to techniques helps to look at the teacher's role in the classroom use of tools.

5. LINKING CAS AIDED PATTERN DISCOVERY AND PAPER-AND-PENCIL TECHNIQUES

Since CAS first appeared in classrooms, there has been a recurrent debate about what should happen to paper-and-pencil techniques. Authors who see these techniques simply as skills tend to think that, if there is some necessity for students to learn them, this learning should be 'resequenced' as late as possible in order to avoid interference with conceptualization (§ 9). Other authors refer to paper-and-pencil proficiencies as sometimes valuable and meaningful. They recognize that technology changes the scene and try to identify 'lists of basic skills' that mathematics educators would agree are necessary for students to know how to perform by hand, even in a technological environment.

Goldenberg (2003) wonders whether "algebra is dead" now that "CAS do, with no effort, what we previously thought we wanted the students to do" (ibid., p.13). He reminds us that the role of algebra is not just to solve practical problems. Algebra can play a role in "opening up various black boxes, including the ones we called patterns" and thus "some algebra skills that are no longer needed for *finding* answers still remain essential for *understanding* answers..." (ibid., p.17) He proposes the example of the expansion of $(x-1)(x^7+x^6+x^5+x^4+x^3+x^2+x+1)$ 'collapsing' to produce (x^8-1) CAS gives the answer, but does not give insight into the process involved (ibid., p.29).

Box 5-3.

A challenge: Find the nth order derivative of $(x^2+x+1).e^x$

(Trouche & al 1998)

This situation comes from a booklet made up from reports by 12th grade students on their solutions to a number of 'challenges'. Here, the challenge was: for every positive integer n find the nth order derivative of $(x^2+x+1).e^x$. Two students presented their work. Their first solution appears on the two TI-92 screens below.

Looking for a pattern

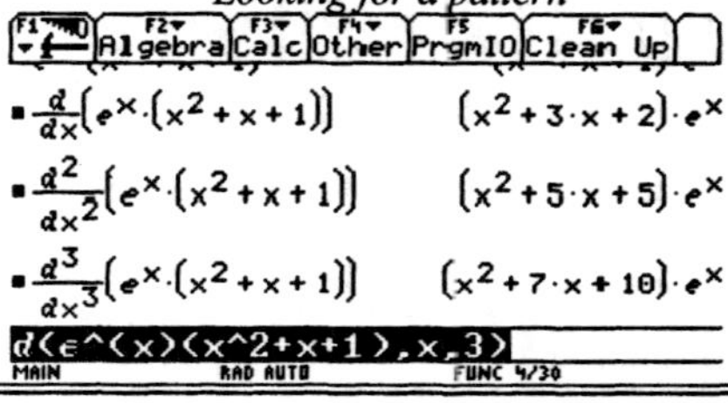

Demonstrating

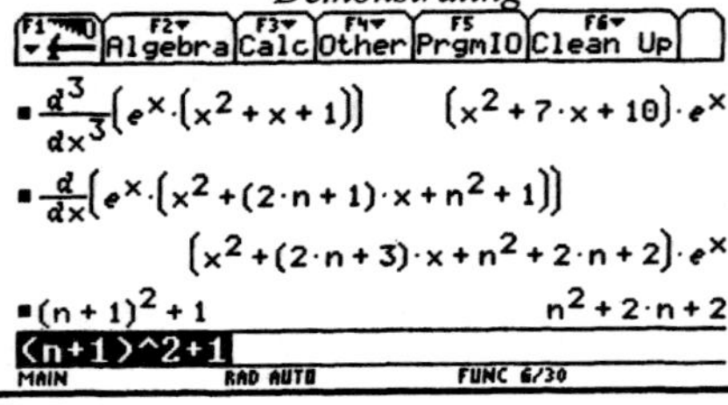

Then the students wrote:

"Using the TI-92, we discovered a pattern and proved it. We had then to look again at this exercise. Actually, we searched for the derivative of a product of two functions u and v, with $u(x)=e^x$ *and* $v(x)=x^2+x+1$. *Every derivative of u is u, the first derivative of v is* $v'(x)=2x+1$, *the second is* $v''(x)=2$ *and the other derivatives of v are zero".*

From this, they calculated the first, second and third derivative of *uv*, then they referred to the 'Leibniz formula' and found $(uv)^{(n)}=uv+n.uv'+\frac{n(n-1)}{2}uv''$. From this, they obtained again the expression for the nth order derivative of $(x^2+x+1).e^x$.

We can think of 'discovering patterns' and 'getting insight into patterns' as two activities, one with the use of a tool and the other with paper-and-pencil. The associated techniques make possible different epistemic contributions to the learning of algebra as the example in Box 5-3 shows.

This example deals with differentiation. The algebraic paper-and-pencil techniques for differentiation are developed in secondary mathematics education mainly for their pragmatic utility and they tend to be seen as meaningless skills. Their epistemic contribution to the understanding of algebraic aspects of calculus is nevertheless important. For instance one cannot understand why CAS simplifies the antiderivative of $x^n e^{x^2}$ only for odd numbers n, without some knowledge of the differentiation of products and chain expressions.

We noted above that the pragmatic value of paper-and-pencil techniques is challenged by 'push button' techniques, and that putting their epistemic value to the fore is not obvious. The situation of Box 5-3 can help to make sense of techniques for differentiation of products. The stability of sets of expressions, like the set of products of the exponential with quadratic polynomials is a consequence of the algebraic properties giving these techniques an epistemic value. The CAS technique of pattern discovery helps to conjecture and prove this stability but it hides the underpinning algebraic properties.

Students challenged to 'explain' the stability had to use the algebraic techniques in a different way as compared to the usual paper-and-pencil differentiation. They produced a second non-CAS solution based on properties of the differentiation of the exponential and quadratic expressions and on the Leibniz formula generalizing the product differentiation to the nth order derivative.

The interest of this situation is the following: a CAS technique of pattern discovery helps to find and prove a property but students recognize that this solution 'tells only part of the story'. Actually, pattern discovery gives the property a meaning at a local level and algebraic techniques are a link with general calculus objects (polynomial, exponential, derivatives...) One can then expect from this interrelation of techniques a more general and reflective understanding of algebraic techniques than in the usual paper-and-pencil exercises on differentiation.

More generally, we cannot envisage students doing mathematics only by using CAS. Rather, we envisage a 'CAS assisted' practice intertwining technology and paper-and-pencil. Thus we should think of the use of CAS as calling for an interrelation between new techniques and paper-and-pencil techniques. Goldenberg's reflection and the above example help to illustrate that this interrelation can be mathematically productive as a specific epistemic contribution can be expected from each type of technique.

6. ACCESSING GENERALIZATION THROUGH SYMBOLIC TECHNIQUES

We will now present a situation which is interesting for similar reasons, but in reverse. In the nth order derivative task, CAS techniques gave a local meaning to the solution and traditional paper-and-pencil work provided for a wider sense. In the tank problem situation (Boxes 5-4 and 5-5), access to generalization is provided by CAS. It is primarily a problem of optimization.

Box 5-4.

The concrete tank problem

(Artigue & Lagrange 1999; Chapter 9, § 3)

A man wants to build a tank. The walls and base of the tank are to be made of concrete 20 cm thick, the base is to be a square, and the tank must contain 32 cubic meters. Let x be the horizontal dimension of the side of the inner square, and let h be the inner vertical of the tank, both measured in meters. What should be the values of x and h to use as little concrete as possible?

Solution

The function giving the quantity of concrete simplifies into $\dfrac{25x^4 + 20x^3 + 4x^2 + 3200x + 640}{125x^2}$.

Generally, students do a graphical or numerical study of this function and observe that a minimum seems to appear near the value x = 4.

In a more mathematical approach to the problem, students can compute, with or without the TI-92, the derivative of this function.

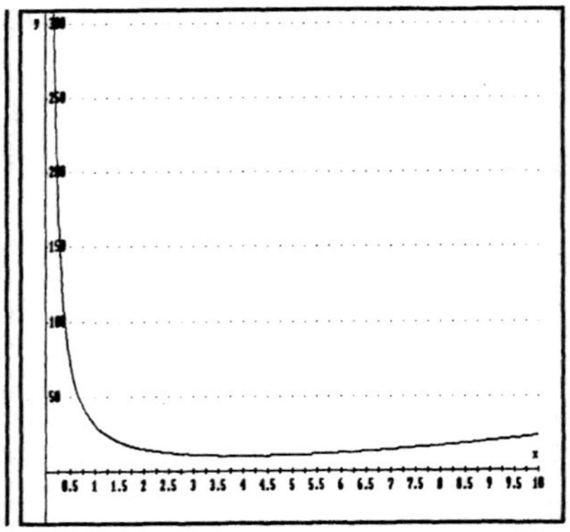

They can find that it has two zeros one for x = 4, and another for x = – 0.4, and generally, they do a numerical study of the sign of this derivative, finding that it is negative between 0 and 4 and positive above 4.

These problems are popular because students can work using precalculus concepts. The graphic and numerical facilities of calculators are excellent supports to encourage students to consider multiple approaches to these problems. In France, students learn about derivatives in eleventh grade and can then solve optimization problems symbolically. They have learned various techniques to tackle these problems and are able to approach calculus concepts when working out these techniques and reflecting on them (Lagrange 2000). By designing lessons for students using a TI-92, we thought that the availability of CAS could help amplify the tasks of optimizing and the associated techniques passing from a particular to a generalized configuration.

Studying a numerical case (Box 5-4) is interesting and the graphical, numerical, and even symbolic capabilities of the TI-92 can help but it has two significant limitations. It primarily 'encourages' the numerical or graphical approach to the problem. Students are not encouraged to use more powerful approaches, such as studying the sign of the derivative. The answer is not remarkable because it does not pave the way for new questions, such as *"Is it a general result that there is a minimum and only one? What can you say of this minimum?"* and so on. Limitations such as those described above do not exist in the generalized problem (Box 5-5).

Box 5-5.

The generalized concrete tank problem

The walls and base of the tank are to be made of concrete of the same thickness e, the base is to be a square, and the tank's inner volume must be V. Let x be the horizontal dimension of the side of the inner square, and let h be the inner vertical of the tank, both measured in meters.

The aim of the problem is to know whether a value of x and h exists that uses as little concrete as possible. In addition, we want to know how this value depends on e and V.

Solution

Students can adapt the function giving the quantity of concrete from the numerical case and get its derivative from the TI-92. Then the task is more difficult because they cannot conduct a graphical study of the function or a numerical study of the derivative as they usually do, but study symbolically its sign by factoring the derivative. The TI-92 gives this factorization $\frac{2e(x+2e)[x-(2V)^{1/3}][x+(2V)^{1/3}+(2V)^{2/3}]}{x^3}$, showing that the derivative has the same sign as $[x-(2V)^{1/3}]$ for positive x.

A minimum quantity of concrete is used at $x = (2V)^{1/3}$. The problem was trialed during a TI-92 experiment with 11th grade students and Chapter 9 (§ 3) will report on its place in the experiment and on students' work. We discuss here a possible student solution. Solving the numerical case should help students to understand the problem and try a numerical or graphical approach. The main difficulty will be to find an algebraic expression of the concrete volume with respect to x. With the generalized problem, students will meet the limitations of their graphic calculator techniques. They will be driven towards a symbolic technique that they learned before, but do not use in numerical cases when they see more sense in graphic and numerical approaches. They will be able to perform this technique only with the help of the TI-92 not just because the expression is 'big' but really because 11th grade students' knowledge of algebraic manipulation and differentiation is too weak to handle parameterized expressions.

We expected that students could answer the above questions in the following terms -- *"The minimum depends on V, but it does not depend on e"* -- and see that this issue of dependence and independence is more important than the value of x itself. The general problem is thus not simply a continuation of the numerical problem; it reveals the limitations of an existing technique and promotes a new, more general and symbolic technique. The objects that the general technique handles bring more sense to the problem. It is an example of CAS providing new techniques in interrelating with old techniques, opening a new understanding of optimization. These new techniques are possible because the use of CAS allows students to interpret calculations with symbolic constants, or parameters, as a continuation of the same calculation with explicit values. As another example, Chapter 7 offers a report on students' solutions of two-variable systems and a discussion of the role of parameters and of techniques -- or schemas -- in the CAS context.

The techniques presented in the two last sections are richer than simple 'push button' techniques. Their value follows naturally from the potentialities of computer tools. Easy computation helps pattern discovery. Recovery of a memorized numerical calculation allows students to rethink a technique so as to introduce generalization and make use of CAS for proof. Situations to work on these techniques can then be easily introduced into teaching. However, we are aware that not all obstacles will be suppressed. For instance it is relatively easy to introduce parameters to generalize a numerical situation, but students may have difficulties resulting from the plurality of roles that a letter can play, as we will see in the example of Chapter 7.

7. TECHNIQUES FOR MANAGING EXPRESSIONS

In this section, we return to the techniques for managing expressions; their importance and difficulty were shown above in the situation involving factorization. For instance, when students use a symbolic calculator on an everyday basis, these techniques are a necessity for effectiveness and reflectivity. As an example, students using the TI-92 algebraic window have to learn to consciously use the items of the *Algebra* menu (*Factor*, *Expand*, *ComDenom*), to decide whether expressions are equivalent, and to anticipate the output of a given transformation on a given expression. Since a CAS does not generally check conditions for the validity of a transformation involving for instance radicals or quotients[1], a student must learn what s/he has to control and what s/he can trust in CAS operation.

Box 5-6.

Working on techniques for finding equivalent expressions (11th grade)
(Artigue & Lagrange 1999)

The following screens illustrate three tasks that we proposed for work on the equivalence of expressions. This work is a consolidation of high school algebra *and* an introduction to the TI-92 algebraic techniques.

In the first task, (screen A) students had to enter expressions and observe the TI-92 simplification. They then had to identify the mathematical treatment that the simplification carried out. We chose the expressions (see left side of screen A) to obtain a variety of simplifications (see right side of screen A). Expanding, factoring, reordering, partial fractional expanding, and cancellation by a possibly null expression

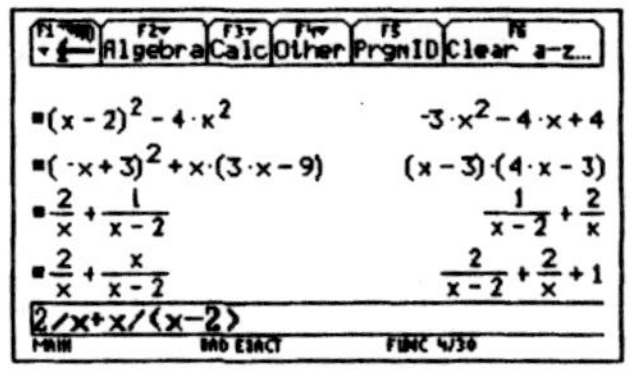

Screen A

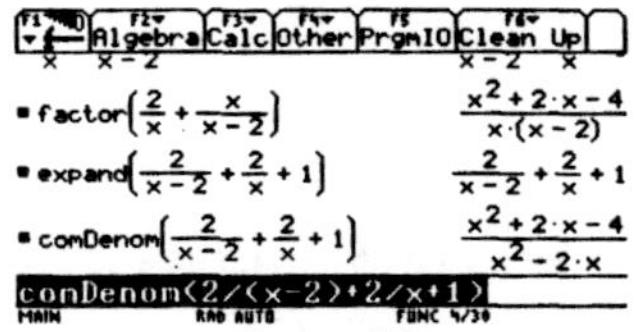

Screen B

In the second task, (screen B) students had to explore the effect of algebra menu items on the same expressions. In this task, they learned to identify the algebraic transformations and their TI-92 syntax. Students also learned how to copy an expression into the entry line. Therefore, they saved time and effort by not entering expressions several times.

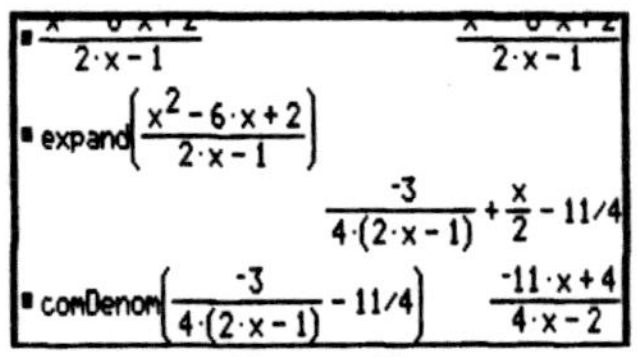

Screen C

In the third task (screens C and D) students had to look for equivalence in four expressions:

$$G = \frac{x^2 - 6x + 2}{2x - 1},\ H = \frac{-11x + 4}{4x - 2} + \frac{x}{2},$$

$$I = \frac{3}{4(2x-1)} - \frac{x}{2} + \frac{11}{4},\ J = \frac{(x + \sqrt{7} - 3)(x - \sqrt{7} - 3)}{2x - 1}.$$

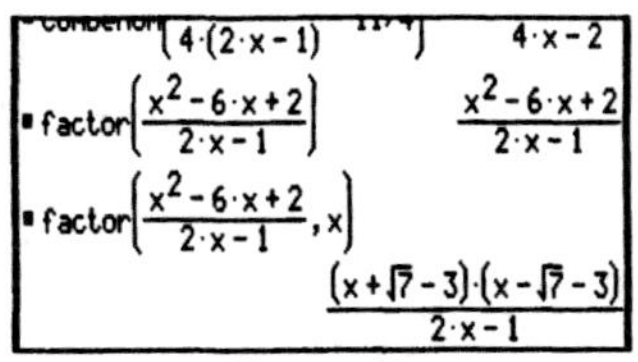

Screen D

To encourage students to use various transformations, we offered expressions in different forms: reduced, more or less expanded, and factorized.

With this aim, we developed a set of three tasks (Box 5-6) to help students acquire flexible use of the TI-92 commands for algebraic transformations useful in working on the equivalence of expressions[2]. This situation was inspired from situations proposed in a 10th grade class in Guin & Delgoulet (1997) which will be analyzed in detail (Chapter 9, § 2). The goal of the first task was to make students aware of the output of the TI-92 simplification and the many possible equivalent forms of an expression. In the second task, our goal was to link the understanding of general forms of expressions with the various items of the Algebra menu. In addition, we wanted students to remember the associated TI-92 syntax. They could also notice that the number of transformed expressions depends on the original expression itself. For instance, every transformation of $\frac{x-2}{x.x-2x}$ gives $\frac{1}{x}$, whereas *Expand*, *ComDenom* and *Factor* have different effects on $\frac{2}{x}+\frac{x}{x-2}$. Interesting discussion may follow this observation. In this task, students may also learn how to copy an expression into the entry line, saving time and effort by not entering expressions several times.

After completing this task, students could learn to use these transformations to decide whether two given expressions are equivalent. With CAS, the technique is as follows: enter the two expressions separated by the equals sign, and simplify. CAS generally returns *true* for equivalent expressions[3]. The epistemic value of this technique is poor because it provides no insight into the reasons for the equivalence. Although the technique is simple, it requires two expressions to be entered and thus can be tedious for complex expressions. We designed the third task to make the use of transformations which are more convenient than entering a test of equality.
The third task gives a rational expression G along with three other apparently equivalent rational expressions H, I, and J. For instance G was $\frac{x^2-6x+2}{2x-1}$, H was $\frac{-11x+4}{4x-2}+\frac{x}{2}$, I was $\frac{3}{4(2x-1)}-\frac{x}{2}+\frac{11}{4}$, and J was $\frac{(x+\sqrt{7}-3)(x-\sqrt{7}-3)}{2x-1}$. With these expressions, a good strategy is to expand G. It yields an expression opposite in sign to I. The user can then copy this expression into the entry line, split the first two terms, and apply the *ComDenom* command. G and H are thus proved equivalent. J is a factored form of H, but the mere Factor command does not transform H into J (Screen B and C). Because J is a 'radical' factorization of H, a special form of the command must be applied. The solution described in Box 5-6 is one of several possible strategies. This topic can prompt many rich mathematical discussions in the classroom.

Pragmatically, these techniques are a necessity for proper use of CAS and their development in the classroom provides opportunities for mathematical discussion. Their epistemic role is clear, as they shed light on the structure and equivalence of algebraic expressions. However, this does not ensure that they will easily gain a place in 'standard' teaching, because of institutional obstacles: the institutional values of the school are defined relative to paper-and-pencil techniques and the dependence of these techniques on a tool is not recognized. We mentioned above that teachers are reluctant to devote time and discussion to techniques that they think too far from 'official' mathematics. They simply reflect the position of the institution: even techniques for managing the graphic window, that would be very useful for students and mathematically meaningful, have no official status in French secondary teaching[4].

8. THE OBSOLESCENCE OF PAPER-AND-PENCIL TECHNIQUES AND TEACHERS' WORK

Many teachers have not yet really considered classroom use of technology[5]. It is an indicator of the difficulties of this endeavor confirmed by the observation of teachers in Chapter 4. The intriguing fact is that even when the introduction of a technology has been well prepared by an epistemological analysis and situations have been proposed, implementation by teachers still looks like a struggle to give birth to a more personal creation. As indeed it is. In our view, new techniques and the way in which they change the teaching and learning of a mathematical domain are not given with the tool and cannot even be just thought of in terms of the design of tool-aided lessons. When a teacher wants to introduce technology, s/he has to integrate these techniques into his/her own understanding of the domain, into his/her own personality and to create relevant situations, certainly not an easy task.

Schneider (1999) offers an example where two teachers wanted to introduce students to TI-92 use in the study of logarithmic functions. They had to entirely rethink their teaching because the techniques they used to work with became obsolete. Without the TI-92, a central task was to solve exponential equations. Students progressively built techniques relevant for a variety of equations and learned about the properties of logarithmic functions by reflecting on these techniques. The teachers became rapidly aware that the TI-92 solved the equations in one easy action and that all had to be rebuilt. The outcome was an entirely new approach to the domain, where symbolic techniques were complemented by graphic and numeric exploration. It is striking that this elaboration appears to be work for the teachers themselves -- or maybe on themselves -- rather than a creation to share with colleagues. A praxeology (Box 5-1) is not just an organization of mathematical contents. At classroom level it offers teachers 'command levers' with which to make students enter the study of a domain. Thus a teacher cannot just receive and apply a new praxeology. S/he has to create something new.

9. THE POTENTIALITIES OF TECHNOLOGY

The conclusion of this chapter is that the potentialities of new tools can only be appreciated by considering the impact of technology on existing techniques and the possible new techniques that students can develop as a bridge between tasks and theories. This is certainly a different viewpoint from that of an influential mathematics-education tradition, which tends to stress an opposition between skills and understanding. In this section we will look at this tradition, see how it lives on in conceptualizations of the use of new tools, and how these conceptualizations converge on the idea of a direct access to concepts, an idea that, from our perspective, cannot really account for the potentialities of technology.

The opposition between manipulation and understanding is ancient especially in the study of algebra. Rachlin (1989) states, "Teachers in the USA even (in 1890) were opposed to what they saw as an overemphasis on manipulative skills and were calling for a meaningful treatment of algebra that would bring about more understanding". In the past fifteen years, the idea that universal access to new technology would "enable us to modify our skill-dominated conception of school algebra and rebalance it in favor of objectives related to understanding and problem solving" (Kieran & Wagner 1989) gained greater acceptance.

To many authors, CAS was an appropriate technology for this "new balance" because it is not limited by the approximate treatment of numbers or by the necessity of programming. Mayes (1997) states that authors of research papers on CAS often study how CAS may help "set a new balance between skills and understanding" or "resequence skills and understanding".

As early as 1988 Heid published a paper about the educational use of CAS. Her guiding hypothesis was that:

> If mathematics instruction were to concentrate on meaning and concepts first, that initial learning would be processed deeply and remembered well. A stable cognitive structure could be formed on which later skill development could build (Heid 1988, p.4).

This paper had a great influence and authors very often refer to it as a confirmation for hypotheses about benefits of technology and especially CAS. One of these authors, Pérez (1998) published a text following a talk at the International Conference on Mathematics Education (ICME 8) and he interpreted Heid's study as a proof of the many advantages of CAS including "students' definite progress toward higher levels of formal thinking and easier integration of conceptual representations".

It is interesting to look in detail at Heid's argumentation. Her research is about an experiment involving the use of early computer programs -- a symbolic calculator software with a command line user interface and a graphing application without connection to the calculator -- in a project involving a new approach to introducing calculus. The author was also the teacher and she chose to delay training in computational skill, to develop graphic approaches to concepts and encourage reflection on the meaning of computer results, and to set students wider classes of problem to solve. She compared her students' proficiency with that of a control group following a 'traditional' curriculum. Delayed skill training did not harm her students and they achieved some more varied representations of concepts.

Using computers, more varied approaches are certainly possible and Heid's experiment provides a remarkable example of such use in a calculus course. Although maybe not this definite progress "towards higher levels of formal thinking", students' tendency towards more varied representations is worth noting. Ruthven (2002) took a closer look at the conditions that made technology contribute to this tendency:

> In the experimental classes, the constitution of a quite different system of techniques appears to have played an important part. The shift to "reasoning in non algebraic modes of representation [which] characterized concept development in the experimental classes" (p.10) not only created new types of task, but encouraged systematic attention to corresponding techniques (...). Not only did the 'conceptual' phase of the experimental course expose students to (...) wider techniques; it also appears to have helped students to develop proficiency in what had become standard tasks, even if they were not officially recognized as such, and had not been framed so algorithmically, taught so directly, or rehearsed so explicitly as those deferred to the final 'skill' phase.
>
> It deferred routinization of the customarily taught skills of symbolic manipulation until the final phase of the applied calculus course, while the attention given to a broader range of problems and representations in the innovative main phase supported development of a richer conceptual system. Equally, however, (...) this conceptual development grew out of new techniques constituted in response to this broader range of tasks, and from greater opportunities for the theoretical elaboration of these techniques. At the same time, standard elements emerged from these new tasks, characteristic of the types of problem posed and the forms of representation employed, creating a new corpus of skills distinct from those officially recognized.

Artigue (2002) has noted that Chevallard's approach gives technique "a wider meaning than is usual in educational discourse" comprising not just recognized routines for standard tasks but more "complex assembl[ies] of reasoning and routine work", whereas mainstream mathematics education research delimits the technical domain more narrowly in terms of *routine manipulations*, *computational procedures* and *algorithmic skills*.

Ruthven's analysis above takes *technique* in this wider meaning and sheds helpful light: the traditional opposition of concepts and skills should be tempered by recognizing a technical dimension in mathematical activity which is not reducible to skills. A cause of misunderstanding is that, at certain moments, a technique can take the form of a skill. This is particularly the case when a certain 'routinization' is necessary. But techniques must not be considered only in their routinized form. In this chapter we tried to show that when CAS is used, the technical dimension is different, but it retains its importance in giving students understanding. The work of constituting techniques in response to tasks, and of theoretical formulation of the questions posed by these techniques remains fundamental to learning.

This chapter provided a first approach to the techniques appearing when new tools are used and a sense of their variety. We have restrained our reflection to the fact that new artifacts were designed as tools to facilitate some techniques and so necessarily have a strong impact on the technical level of mathematical activity, making new techniques possible and old techniques in some sense obsolete. However we have also mentioned that changes in the teaching and learning of a mathematical domain resulting from this impact are not directly determined by an artifact. These changes cannot be appreciated without considering the evolving relationship between users and tools, an idea that the next chapter will develop, stressing the transformation of an artifact into an *instrument* (Chapter 6) for mathematical work. There the term *instrumented techniques* will be used to denote the way in which new techniques are linked to the tool that makes them possible but also to the mathematical domain that they address and to the user's representations of both.

NOTES

---

1. For instance, the TI-92 gives a solution -1 for the equation of a real unknown x: $\sqrt{x}\sqrt{x-3}=-2$.
2. Chapter 9 § 3 will discuss the implementation of this situation within a curriculum.
3. Actually this is true only for expressions belonging to a set where a canonical form for equivalent expressions exists and is implemented in the calculator.
4. In 1998 the French Ministry for Education designed an experimental (non official) baccalaureate. The paper included an interesting question about characteristics of a window to conjecture the intersection of two curves. No change followed in the official exam (Guin & Trouche 2002, p.109).
5. Little data is actually available on the use of technology by teachers and biases can often be suspected. For instance BECTa (2002) maintains that the proportion of upper secondary school pupils in the UK never -- or hardly ever -- using ICT in their mathematics lessons is as much as 82%. This statistic however ignores the extent to which graphic calculators are used, since the survey in question appears not to have classed these as ICT.

REFERENCES

Artigue M. (2002) Learning mathematics in a CAS environment: The genesis of a reflection about instrumentation and the dialectics between technical and conceptual work, *International Journal of Computers for Mathematical Learning* 7(3), 245-274.

Artigue M. & Lagrange J.B. (1999) Instrumentation et écologie didactique de calculatrices complexes: éléments d'analyse à partir d'une expérimentation en classe de Première S, in D. Guin (Ed.), *Actes du congrès Calculatrices symboliques et géométriques dans l'enseignement des mathématiques* (pp. 15-38). Montpellier: IREM, Université Montpellier II.

BECTa (2002) *ImpaCT2: The Impact of Information and Communication Technologies on Pupil Learning and Attainment.* London: Department for Education and Skills. (http://www.becta.org.uk/research/index.cfm.).

Chevallard Y. (1999) L'analyse des pratiques enseignantes en théorie anthropologique du didactique, *Recherches en Didactique des Mathématiques* 19(2), 221-266.

Goldenberg P. (2003) Algebra and Computer Algebra, in Fey (Ed.), *Computer Algebra Systems in Secondary School Mathematics Education*, NCTM monograph.

Guin D. & Delgoulet J. (1997) *Etude des modes d'appropriation de calculatrices graphiques et symboliques dans une classe de seconde.* Montpellier: IREM, Université Montpellier II.

Guin D. & Trouche L. (2002) *Calculatrices symboliques. Transformer un outil en un instrument du travail mathématique: un problème didactique.* Grenoble: La Pensée Sauvage éditions.

Heid M.K. (1988) Resequencing skills and concepts in applied calculus, *Journal for Research in Mathematics Education* 19(1) 3-25.

Kieran C. & Wagner S. (1989) The research agenda conference on algebra: background and issues, In Kieran & Wagner (Eds.), *Research Issues in the Learning and Teaching of Algebra* (pp. 1-10). NCTM-LEA.

Lagrange J.B. (2000) L'intégration d'instruments informatiques dans l'enseignement: une approche par les techniques, *Educational Studies in Mathematics* 43(1), 1-30.

Mayes R. (1997) Current state of research into CAS in mathematics education, in J. Berry, J. Monaghan (Eds.), *The state of computer algebra in mathematics education* (pp. 171-189). Chartwell-Bratt.

Monaghan J., Sun S. & Tall D. (1994) Construction of the Limit Concept with a Computer Algebra System, in *Proceedings of PME 8* (Vol. 3, pp. 279-189). Lisbon: University of Lisbon.

Mounier G. & Aldon G. (1996) A Problem Story: Factorisations of $x^n - 1$, *International DERIVE Journal* 3(3), 51-61.

Pérez Fernández J. (1998) Los systemas de cálculo simbólico en la enseñanza de las matemáticas, in Hodgson & al (Eds.), *Selected lectures 8th International Congress on Mathematical Education* (pp. 345-368). SAEM THALES.

Rachlin S. (1989) The research agenda in algebra: a curriculum development perspective, in Kieran and Wagner (Eds.), *Research Issues in the Learning and Teaching of Algebra* (pp. 257-265). NCTM-LEA.

Ruthven K. (2002) Instrumenting mathematical activity: Reflections on key studies of the educational use of computer algebra systems, *International Journal of Computers for Mathematical Learning* 7(3), 275-291.

Schneider E. (1999) La TI-92 dans l'enseignement des mathématiques, des enseignant(e)s découvrent la didactique des mathématiques, in D. Guin (Ed.), *Actes du congrès*

Calculatrices symboliques et géométriques dans l'enseignement des mathématiques, Mai 1998 (pp. 49-60). Montpellier: IREM, Université Montpellier II.

Sierpinska A. & Lerman S. (1996) Epistemologies of mathematics and mathematics education, in Bishop & al (Eds.), *International Handbook of Mathematics Education* (pp. 827-876). Dordrecht: Kluwer Academic Publishers.

Trouche L. & al (1998) *Faire des Mathématiques au lycée avec des calculatrices symboliques*. Montpellier: IREM, Université Montpellier II.

Chapter 6

AN INSTRUMENTAL APPROACH TO MATHEMATICS LEARNING IN SYMBOLIC CALCULATOR ENVIRONMENTS

Luc Trouche
LIRDEF, LIRMM & IREM
Université Montpellier II, France
trouche@math.univ-montp2.fr

Abstract: A rapid technological evolution (Chapter 1), linked to profound changes within the professional field of mathematics (Chapter 3), brings into question the place of techniques in mathematics teaching (Chapter 5). These changes have created serious difficulties for teachers; obliged to question their professional practices, they make different choices regarding integration of new technologies and techniques (Chapter 4), choices that are linked to their mathematical conceptions and to their teaching styles.

In this chapter, we place ourselves on the side of the students. We have already seen (Chapter 1) that they seem to adopt the new computing tools faster than the institution. In this chapter, we study more precisely their learning processes related to their use of symbolic calculators.

First of all, we pinpoint the didactic phenomena taking place in the experiments; subsequently, we suggest a new theoretical approach aimed at giving a better description, for each student, of the transformation of a technical tool into an instrument for mathematical work.

Key words: Computational transposition, Instrumentation and instrumentalization process, Instrumented technique, Operational invariants, Schemes.

1. DIDACTIC PHENOMENA APPEARING IN FIRST EXPERIMENTS

Research studies on symbolic calculators as environments have been conducted in France since 1995. These were preceded by research on calculator integration (from 1980), and by research on DERIVE software integration (institutionally supported in France from 1991). These studies have revealed many didactic phenomena. Artigue (1997) distinguishes two interrelated classes of phenomena: those linked to knowledge transposition and those linked to students' adaptation to new environments.

1.1 Didactic phenomena linked to processes of knowledge transposition

These processes are linked to *computational transposition* (Box 6-1), described by Balacheff (1994) as "work on knowledge which offers a symbolic representation and the implementation of this representation on a computer-based device".

Artigue (1997) brings out two phenomena linked to these processes:

- the phenomenon of *pseudo-transparency*, linked to the gap between what a student writes on the keyboard and what appears on the screen (a gap arising from differences between two representation modes, internal and interface):

> [To enter (a+2)/5], some students, having correctly written a couple of parentheses around (a+2), are surprised to see a screen display without parentheses. They wonder if their production is correct, or not. Parentheses appearing and disappearing seems to be a mysterious game they can't understand, precisely because they have not mastered parentheses techniques.

- The phenomenon of *double reference*, linked to the double interpretation of tasks, depending on the work environment (paper-and-pencil or computerized). Artigue (ibid.) evokes in the following terms the rational factorization of $x^n - 1$ in a 11th grade class, with DERIVE software (Box 5-2):

> In a paper-and-pencil environment, polynomial factorizations are linked, at this school level, to the search for real roots (...). In the software environment, these rational factorizations come first from factorizations in Z/pZ (...): the factorization by (x – 1) for example is obtained only if n is a prime number and the factorization by (x – 1) (x + 1) only for n = 4 (...). Students choosing the machine interpretation have much greater difficulties in producing conjectures.

Box 6-1.

Computational transposition
(Balacheff 1994)

Balacheff defined *computational transposition* in the following terms:
"A representation of the world is not the world itself. This now largely shared assertion can be taken as a commonplace. Nevertheless, ... to understand it, we have to go further and consider that a representation is not an approximation (i.e. a simplification) of its object in order to re-present it. Each representation has properties which come both from modeling choices and from chosen semiotic modes. These properties have, a priori, no connection with the represented world. Moreover, as a material device, a computer imposes a set of constraints which themselves will impose an appropriate transformation allowing the implementation of the adapted representation.
I will name as *computational transposition* this work on knowledge which offers a symbolic representation and the implementation of this representation on a computer-based device, in order to *show* knowledge or to manipulate it. In a learning context, this transposition is particularly important. It implies indeed a contextualization of knowledge, with possible important consequences for learning processes".
Balacheff distinguishes constraints linked to the internal universe of a machine (for example, the program for representing a circle, Figure 6-1) from interface constraints (for example, the screen representation of a circle distorted by pixellation).

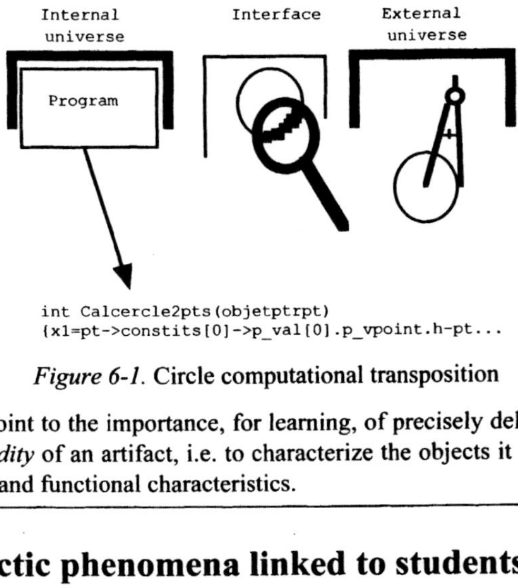

Figure 6-1. Circle computational transposition

These phenomena point to the importance, for learning, of precisely delimiting the domain of *epistemological validity* of an artifact, i.e. to characterize the objects it gives access to and to identify its semiotic and functional characteristics.

1.2 Didactic phenomena linked to students' processes of adaptation to environments

Experiments also reveal processes of adaptation to environments: in these processes, the constraints and potentialities of computerized environments play a determining role.

1.2.1 Perceptual adaptation processes

First of all these processes are linked to the potentialities of calculators as regards *visualization*, in terms of graphical as well as algebraic *representatives* (Schwarz & Dreyfus 1995). The influence of 'direct'[1] perception is most widely noted in the graphic frame. We showed (Guin & Trouche 1999), for example, that answers to the question "Does the function f, defined by $f(x) = \ln x + 10 \sin x$, have an infinite limit as x tends to $+\infty$?" depend strongly on the working environment (even though elementary theorems make it possible to answer *yes* to this question).

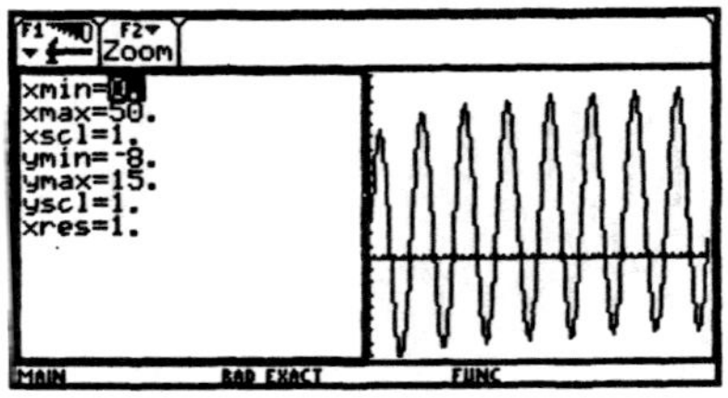

Figure 6-2. A graphic representation of the function f: $x \mapsto \ln x + 10\sin x$

In a graphic calculator environment, 25% of students answered *no*, appealing to the oscillation of the observed graphic representation (Figure 6-2); in a paper-and-pencil environment, only 5% of students answered *no*.

The importance of framing perception in algebraic terms was pointed out by Artigue (1997); students, in a DERIVE environment, had to explain how to move from the equation $2x - 5y = 8$ to the equation $12x - 30y = 48$, then $6x - 15y = 24$.

> On first passage, the teacher gave an indication: multiply by 6. This will favor a solution based upon formal analogies which are essentially perceptual (...). [Students] suppose that 6 has been obtained by dividing, but wonder how this division could be done. After some hesitation, they decide to try using DERIVE, to do something which according to them, "*would probably give nothing, but trying costs nothing*": they enter the two equations under division and ask DERIVE to simplify. DERIVE answers with: 3 = 3, which amuses students ("*it's trivial!*") but also intrigues them. They do not try to understand this answer, but decide to do the same thing with the first equation. This time, $(12x - 30y = 48) / (2x - 5y = 8)$ gives 6 = 6.
>
> As DERIVE answers 6 = 6, when the teacher's answer is 6, so, in the second case, when DERIVE answers 3 = 3, the right answer must be 3.

Thus, perceptual adaptations can come into play within both graphical and algebraic frames, but this does not guarantee the establishment of relationships between these two frames: Dagher (1996) shows that frequent use of software allowing algebraic and graphical representations of functions to be manipulated does not necessarily help students to build an efficient articulation between these two frames.

Perceptual adaptations are also linked to potentialities for *animation*. We have pinpointed numerous manifestations of this in calculator environments; for example (Trouche 1995), students had to find a parabola tangential to three given lines (Figure 6-3). To perform this task, students tested diverse parabola equations. In order to check if their parabola was a correct solution, they zoomed in on the contact point between a line and the parabola. They supposed this contact to be 'good' if, after several zooms, curve and line appeared confounded on the calculator screen. At the end of this work, the teacher asked: "*How can a tangent to a curve be defined?*"

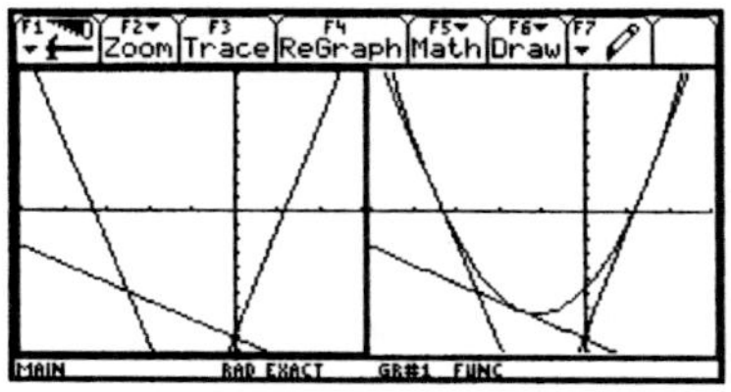

Figure 6-3. Search for a parabola tangential to three given lines

The first answer proposed from the class was: *"a line is the more tangential to a curve the more common points it has with it"*. This definition does not correspond with any taught knowledge acknowledged by the institution; it is the simple translation of students' observation of a 'good' contact between a curve and a line on a calculator screen composed of pixels (Figure 6-3). This is a visualization effect, linked to the computational transposition (particularly, here, the constraints of discrete traces, Box 6-2).

In tackling the same task, the strategy of finding by trial and error a curve of equation $y = ax^2+bx+c$ also leads to the construction of knowledge (related to the roles of the coefficients a, b and c). In this context, students' activity is essentially based on observing the displacement of curves through modification of coefficients in their equation. Thus, for c, students claim: "*when c increases the parabola goes up, when c decreases the parabola goes down*". Even after a clarification from the teacher ("*c is the y co-ordinate of the intersection of the curve with the y axis*"), it is often the first interpretation that is memorized: when the teacher asks about the sign of the coefficient c in the equation of the specified parabola (Figure 6-4), some students answer that c is equal to zero in the case of the left parabola ("*the parabola is at the level zero*") and c is negative in the case of the right parabola ("*the parabola is underneath the x-axis*"), although c is strictly positive in both cases.

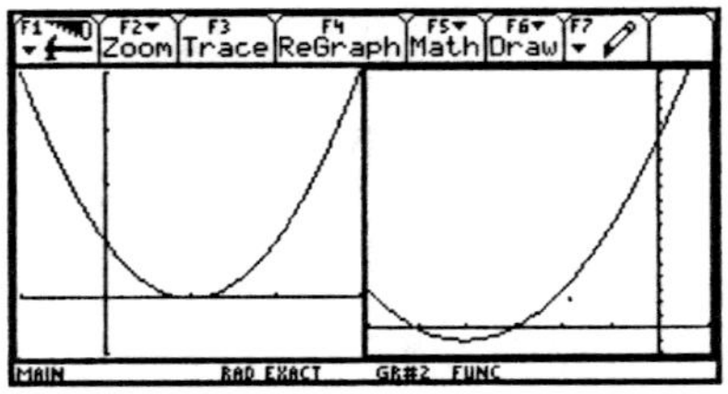

Figure 6-4. Two unknown parabola equations

This reaction shows the importance of movement for students' perception and description of what appears on a screen (what students remember is more the *change* between two images, rather than the properties of each image. The same reasoning in terms of animation (more precisely, the possibility of moving a point on a curve thanks to the *Trace* command) could explain students' conception of a function graph, in a calculator environment, as the trajectory of a moving point, rather than as a set of points whose coordinates are (x; f(x)).

Box 6-2.

Constraints of discrete plots and some consequences
(Guin & Trouche 1999)

Many phenomena arising in relation to the graphical interface of calculators are linked to the presentation of discrete plots on a screen composed of a finite number of pixels.

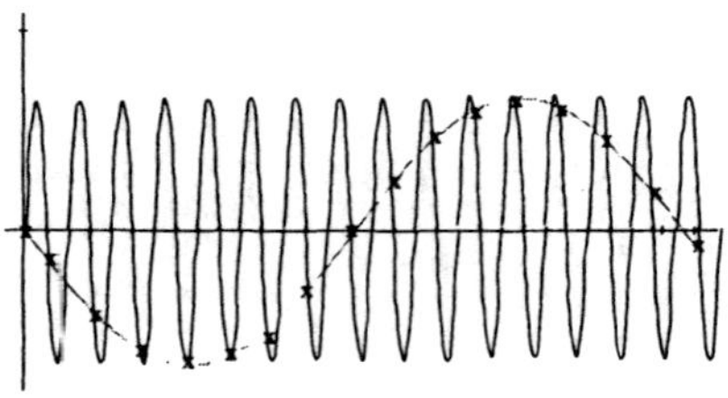

Figure 6-5. A diagram showing the consequence of a discrete plot: an usual period for sine function

For example, if the represented function is periodic (sine function, Figure 6-5) and if the distance between two computed points is close to its period, the two computed images will also be close together. As the calculator joins these two points, the oscillation between these points will not be shown.

1.2.2 Phenomena linked to the organization of students' work

The *multiplicity* of *easily available* commands has effects on the economy of students' work.

Students carry out trials and tests without paying attention to their organization and control. They hope that, within a reasonable time, they will obtain something interesting.

> Observations show that these fishing behaviors can be productive for students, often more productive than more reflective behaviors available to them. The low cost of these trials and their productivity tend to discourage retroactive approaches, involving looking back and modifying accordingly, generally considered as essential in generating the cognitive adaptations hoped for (Artigue 1997).

In the same category, Defouad (2000) pinpoints a *zapping* phenomenon (which consists of quickly changing graph window, without having time to analyze each of the representations obtained), an *oscillation* phenomenon (students oscillating between several techniques and strategies) and an *over-checking* phenomenon (students carrying out multiple checks, using all the means provided by the calculator).

We have also noted similar phenomena, in a calculator environment (Trouche 1997):

- a phenomenon of *automatic transportation*: students enter all the problem data into the calculator, and then look for the command which could give the solution directly:

> [A student] studies a positive sequence u(n) converging toward 0. He wants to determine the value of n from which u(n) will be smaller than 10^{-10}. He takes his calculator, enters u(n) in the sequence editor, enters 10^{-10} in the window setting, as "nmax". Then he wonders what is the right key he has to press in order to have the result?

- a phenomenon of *localized determination*, linked to the difficulty of moving from one *register* to another one (Duval 2000; Guin & Trouche 2002, p.158) and of changing application on a symbolic calculator. It consists of repeating the same type of technique, within the same calculator application, making some adjustments, even if this type of technique does not appear relevant.

> To answer the question "Are there some power functions with curves tangential to the curve of the exponential function?", some students tried in succession (x^2, then x^3, then $x^{2.1}$, then $x^{2.2}$...). Each curve was tested through successive zooms. During the whole activity (taking one hour), the same type of approach was repeated.

While looking back is exploited effectively (unlike in fishing behavior), work remains confined to a single graphical application, with a double consequence:

- firstly, for problem solving: doing mathematics often requires changing one's point of view. Keeping the same point of view, using a single technique, often does not allow a given problem to be solved;
- secondly, for building knowledge: working in a single register, representing a mathematical object in a single form, does not make it possible to form a complete notion of this object.

These phenomena generally appeared in long-term experiments, where students had calculators at their disposal (both at school and at home). These parameters are probably important, facilitating the appropriation of the calculator by students and the stabilization of techniques to perform given tasks. This necessity of taking into account the potentialities and constraints of new tools led us to study appropriation and utilization processes. More generally our interest in *mediation* linked to the learning process (Vygotsky 1962) led us to seek new theoretical approaches, which would yield better understanding of the role of material and symbolic *instruments* within mathematical activity.

2. A NEW APPROACH IN ORDER TO UNDERSTAND AND DESCRIBE NEW PHENOMENA

Recent work in the field of cognitive ergonomics has provided theoretical tools allowing a better understanding of processes of appropriation of complex calculators. Verillon and Rabardel, dealing with training in general (1995) propose a new approach, which essentially distinguishes an *artifact* from an *instrument*:

- an artifact is a material or abstract object, aiming to sustain human activity in performing a type of task (a calculator is an artifact, an algorithm for solving quadratic equations is an artifact); it is *given* to a subject;
- an instrument is what the subject *builds* from the artifact.

This building (Figure 6-6), the so called *instrumental genesis*, is a complex process, linked to characteristics of the artifact (its potentialities and constraints) and to the subject's activity, her/his knowledge and former work methods.

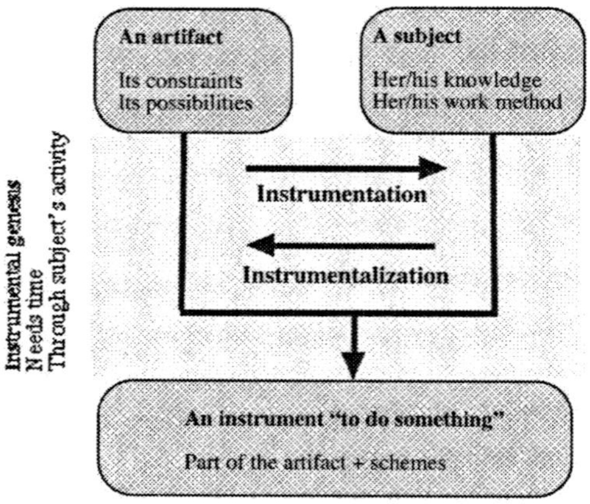

Figure 6-6. From artifact to instrument

This schema rests on some fundamental ideas:

- an artifact *partially* prescribes the user's activity, through its constraints and potentialities;
- instrumental genesis is a process (therefore *needs time*) and has two components, the first one (*instrumentalization*) directed toward the artifact, the second one (*instrumentation*) directed toward the subject;
- a subject builds an instrument *in order to perform a type of task*; this instrument is thus composed of both artifact (actually *a part* of the artifact used to perform these tasks) and subject's *schemes* (Box 6-4) allowing her/him to perform tasks and control her/his activity.

We are going now to make these ideas more precise in the context of symbolic calculators.

2.1 Analyzing constraints and potentialities of symbolic calculators

Computational transposition and design choices produce constraints in a symbolic calculator which Balacheff classifies as internal constraints and interface constraints (Box 6-1).

Regarding general relationships with artifacts, Rabardel (1995) distinguishes three types of constraint: *existence mode constraints*, linked to properties of the artifact as a cognitive or material object, *finalization constraints*, linked to objects it can act on and to transformations it can carry out, and, lastly, *action prestructuration constraints*, linked to prestructuration of the user's action.

Concerning symbolic calculators, we have used (Trouche 1997) both Balacheff's and Rabardel's typologies, distinguishing *internal constraints* (identified as existence mode constraints), *command constraints* (linked to the existence and the nature of specific commands) and *organization constraints* (linked to ergonomic questions, particularly keyboard and menu organization).

Defouad (2000) notes some shortcomings in this typology:

- internal constraints do not cover all existence mode constraints (for example, the nature of the calculator screen is not an internal constraint, but an existence mode constraint);
- all the constraints actually prestructure the user's activity (and not only organization constraints);
- this typology does not take into account various *information levels*: information introduced by the user at the interface, information accessible at the interface, but not open to transformation by the user, and information not accessible at the interface;

- it does not take account of *syntax constraints*, even though these can be decisive when introducing information at the interface.

Box 6-3.

Internal constraints of one graphic calculator

(Bernard & al 1998)

The authors studied internal constraints of the TI-82 calculator. Figures 6-7 and 6-8 show one illustration, linked to implemented algorithms for approximate computation: while the limit at 0 of the given function is 1/6, the table of values and graph of the function give first a value close to 1/6, then, as x approaches 0, produce some oscillations, and finally seem to give, as the function limit, the value 0.

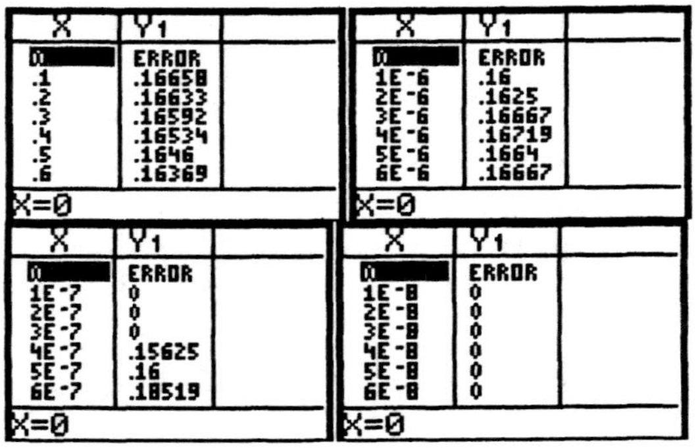

Figure 6-7. Numerical observation of the function f: $x \mapsto \dfrac{x - \sin x}{x^3}$ near 0

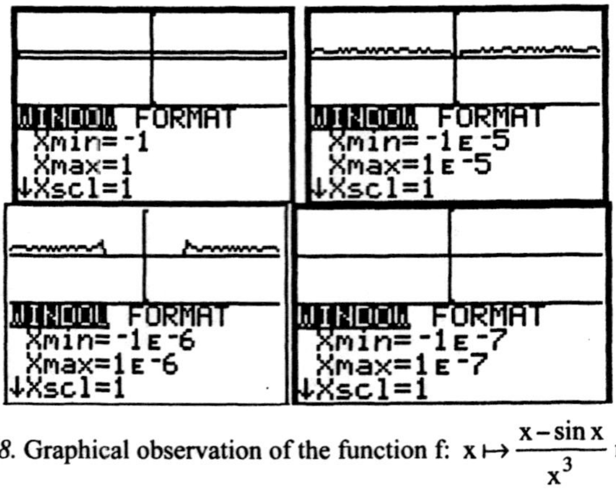

Figure 6-8. Graphical observation of the function f: $x \mapsto \dfrac{x - \sin x}{x^3}$ near 0

Taking account of these remarks, we make precise three types of constraint, all serving to prestructure the user's action and related to a type of task:

- *internal constraints* (in the sense of physical and electronic constraints) intrinsically linked to material. They are linked to information which the user cannot modify, whether accessible or not. They strictly shape action. They include for example processor characteristics, memory capacity (Box 6-3) and screen structure, composed of a finite number of pixels (Box 6-2);

- *command constraints* linked to the various commands in existence and their form (including syntax). They are linked to information accessible at the interface which the user can sometimes modify:

Example 1: the *Range* application of the calculator allows the viewing window to be fitted to a graphic representation of the function. The choices open to the user are relatively free: s/he can choose Xmin and Xmax (but not Xmax smaller than Xmin). The graphic representation of the function which is obtained through setting these values provides feedback allowing a better fitting window to be found.

Example 2: some calculators (Texas Instruments symbolic calculators) require the use of parentheses when computing function values (sin(2), log(3), etc.). This is not the case for other calculators (Casio symbolic calculators) which accept entries such as sin2, or log3: these different design choices can have consequences for students' conceptions about functions.

- last, *organization constraints* linked to keyboard and screen organization, i.e. to available information and command structure.

Example 1: designer choices related to functions (the naming of commands, means of accessing them and their placing within a menu) give a particular point of view on available objects (Appendix 6-1). These choices are linked to an ergonomic study of users' needs, and, at the same time, they favor a particular form of tool use.

Example 2: the placing of the symbol "∞" is not neutral. On Texas Instruments calculators, this symbol is directly given by a keystroke (and it can be manipulated as a number or a letter). On Casio symbolic calculators, it is available only in the CAS application. These different approaches can instill different relationships with this symbol ∞, and, beyond, with the notion itself (Appendix 6-1).

It is possible to discuss the placing of a given constraint into one of the three defined types. But this interest in typology is not strictly in partitioning constraints; it is rather in making easier, for teacher as well as for researcher, an *a priori* analysis of different ways proposed for performing tasks with an artifact. Distinguishing these three levels allows this analysis to be organized in *a given mathematical context* (Box 8-5, for such an analysis of limit computation). Particularly, distinguishing an elementary level of *command* constraints and a more complex level of *organization* constraints permits a distinction, within students' activity, between a level of *gesture* and a level of *technique*.

Analyzing calculator constraints shows clearly that it presents mathematical knowledge in a particular way: "These tools wrap up some of the mathematical ontology of the environment and form part of the web of ideas and actions embedded in it" (Noss & Hoyles 1996). A user is thus not 'free' to use, as s/he wants, a given tool: "This use is, relatively, prestructured by the tool" (Luengo & Balacheff 1998). These constraints do not necessarily lead to impoverishment of activity: by taking in charge part of the work, by favoring exploration in various registers (Yerushalmy 1997), tools open new ways for conceptualization. It is indeed difficult to separate potentialities on the one hand and constraints on the other: both are intimately linked, each facility offered presses the user to realize one type of gesture rather than another.

2.2 Understanding two components of instrumental genesis: one directed toward the artifact, the other directed toward the subject

Instrumental genesis (Figure 6-6) is a process of building an instrument from an artifact. It has two closely interconnected components:

- the *instrumentalization* process, directed toward the artifact;
- the *instrumentation* process, directed toward the subject.

The instrumentalization process, directed by the subject, involves several stages: a stage of *discovery* and *selection* of the relevant keys, a stage of *personalization* (one fits the tool to one's hand) and a stage of *transformation* of the tool, sometimes in directions unplanned by the designer: modification of the tool bar, creation of keyboard shortcuts, storage of game programs, automatic execution of some tasks (the web sites of calculator manufacturers or the personal web sites of particularly active users often offer programs for functions, methods and ways of solving particular classes of equations etc.). Instrumentalization is a process of differentiation as regards the artifacts themselves:

- differentiation regarding the calculator's contents: in making comparisons between students' calculators it is possible to identify differences (from both quantitative and qualitative points of view) between the various programs stored;
- differentiation regarding that part of the artifact mobilized by the subject (for some students, a very small part of calculator, for others a large one).

Instrumentalization is the expression of a subject's specific activity: what a user thinks the tool was designed for and how it should be used: *the elaboration of an instrument takes place in its use.*

Instrumentalization can thus lead to enrichment of an artifact, or to its impoverishment.

Instrumentation is a process through which the constraints and potentialities of an artifact *shape* the subject. As Noss & Hoyles (1996) note: "Far from investing the world with his vision, the computer user is mastered by his tools". This process goes on through the emergence and evolution of schemes (Box 6-4) while performing tasks of a given type. We will study an example of such processes in the following section. As instrumentalization processes, instrumentation can go through different stages. Defouad (2000) analyzes these processes of evolution for students who, after using graphic calculators, then move on to use symbolic calculators (TI-92). He distinguishes two main phases, first an *explosion* phase and second, a *purification* phase:

> At the beginning of instrumental genesis, the student's work seems to be at a crossroads, as if s/he was looking for an equilibrium between her/his former techniques and strategies (linked to graphic calculators) and various new possibilities opened up by TI-92 calculators and the evolution of classroom knowledge. We call this phase an *explosion phase*, as new strategies and techniques appear to burst out; it seems to be characterized by *oscillation*, *zapping* or *over verification* phenomena (§ 1).
> Progressively, students enter into a second phase we call a *purification phase*, where machine use tends to an equilibrium, in the sense of the stabilization of instrumented strategies and techniques. This phase often goes with a fixation on a few commands (and such choices could be different, according to each student).

Box 6-4.

Schemes and conceptualization
(Vergnaud 1996)

Vergnaud distinguishes:
- Conceptions: "one can express them by sequences of statements whose elements are objects, monadic or polyadic predicates, transformations, conditions, circumstances, forms"...
- Competencies: "one can express them by actions judged adequate for the treatment of situations".

He introduces the *scheme* concept, allowing relationships to be established between conceptions and competencies. A scheme is an invariant organization of activity for a given class of situations. It has an intention and a goal and constitutes a functional dynamic entity. In order to understand its function and dynamic, one has to take into account its components as a whole: goal and subgoals, anticipations, rules of action, of gathering information and exercising control, *operational invariants* and possibilities of inference within the situation.
Vergnaud names as *operational invariants* the implicit knowledge contained within schemes: *concepts-in-action* are concepts implicitly believed to be relevant, and *theorems-in-action* are propositions believed to be true. He distinguishes theorems-in-action and concepts-in-action ("truth is not the same thing as relevance"), but insists on their deep links ("theorems-in-action cannot exist without concepts-in-action, as theorems cannot exist without concepts, and vice-versa"). These operational invariants occupy a central place in this frame as: "two schemes are different as soon as they contain different operational invariants".

To better understand the complexity of these two processes, let us make two elements precise:

i) Instrumental genesis is a process of building an instrument, from an artifact, by a subject. This instrument is built from a part of the initial artifact (modified through the instrumentalization process) and through schemes built in order to perform *a type of task* (Box 5-1). A complex artifact such as a symbolic calculator will thus give birth, for a given student, to *a set of instruments* (for example an instrument for solving equations, an instrument for studying function variation, etc.). The articulation of this set is a complex task (Chapter 8, § 2).

ii) Instrumental geneses have both *individual* and *social* aspects. The balance between these two aspects depends on:

- material factors (it is quite obvious that the 'intimacy' of calculator screens favors individual work whereas computer screens allow common work by small student groups);
- the availability of artifacts (sometimes, they are available only at school, sometimes they are lent for the whole school year, sometimes they are students' property);
- the way in which the teacher takes these artifacts into account (Chapter 8, § 2).

Moreover artifacts are mediators of human activity and activity mediated by instruments is always *situated* (Chapter 8, § 2).

Chacon & Soto-Johnson (1998) analyze some effects of these variables on students' behaviors and on their relationships with artifacts: when calculators or computers are available only from time to time, students often develop a critical attitude toward technology; indeed they are sometimes quite confused (because learning in the two environments -- computerized and 'classical' -- is not the same) and frustrated (computers are not available outside laboratory scheduled work).

2.3 Understanding different levels and different functions of instrumented action schemes

Rabardel (1995) introduced the notion of the *utilization scheme* of an artifact to describe a scheme operative within activity mediated by an artifact and distinguishes two such sorts of schemes:

- *usage schemes*, "oriented toward the secondary tasks corresponding to actions and specific activities directly linked to the artifact";
- *instrumented action schemes*, whose "significance is given by the global act aiming to carry out transformations on the object of activity".

All are partially *social schemes*, as their emergence comes, in part, from a collective process involving artifact users and designers. Schemes of usage and instrumented action are deeply linked. A scheme of instrumented action aims to perform a given task. It includes *operational invariants* (Box 6-4).

One can consider instrumented action schemes as a set of usage schemes. Understanding the function of a usage scheme requires it to be considered not in isolation, but as a component of an instrumented action scheme involved in performing a given task.

2.3.1 Usage schemes and gestures

We define a *gesture* as the observable part of a usage scheme. For example, we illustrate (Trouche 2000) the importance of a particular gesture, *approximate detour;* it consists of a combination of keystrokes which results, when working on a symbolic calculator in exact mode, in an approximate value of a symbolic expression. It is not a simple gesture, only oriented toward calculator management: beyond (or psychologically underneath) this gesture, there is a usage scheme, with associated knowledge. Looking for this knowledge involves considering the gesture not as an isolated act but as integrated within an instrumented action scheme employed by the student in order to resolve given tasks.

We identified (Trouche 1996) the three main schemes of instrumented action in which approximate detour appears as those of solving equations, computing integrals, and computing limits.

The observation of students' work shows rules of action, of gathering information, of exercising control (Box 6-4):

- for some students, the approximate detour has always a *determination function* (the approximate value obtained is considered as *the* value);
- for other students it has always an *anticipation or checking* function (obtaining an approximate value may be a step in the process of seeking an exact value).

In other words, approximate detour contributes to building different kinds of knowledge about, say, the real numbers.

2.3.2 Instrumented action schemes and instrumented techniques

One can describe human activity (and students' activity in particular) in terms of *techniques* (Box 5-1), i.e. sets of gestures realized by a subject in order to perform a given task. When a technique integrates one or several artifacts, we will speak of an *instrumented technique.* Instrumented technique is thus the observable part of an instrumented action scheme. For example, an instrumented technique which can be *described* in this way (Trouche 2001) is one for limit computation, in a symbolic calculator environment, as presented by a teacher (Figure 6-9).

Its presentation is made as a tree. In general, such a tree can be:

- more or less 'vast' (in the sense of the number of calculator applications used, of the number of frames evoked, etc.). We can observe in this case that the numerical frame is not used (for numerical observations, for example);
- more or less dense (in the sense of the *granularity* of prescribed gestures). In this case, use of the calculator in order to split the problem is not indicated.

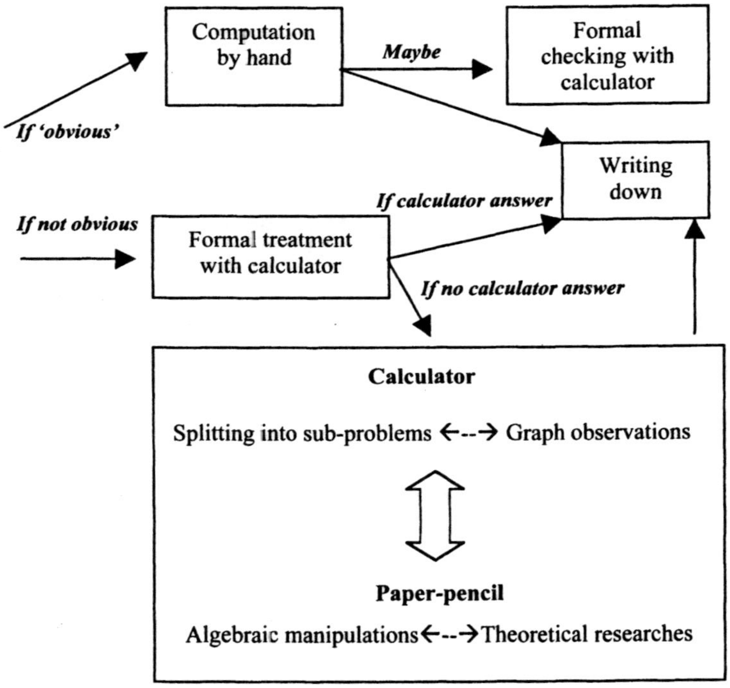

Figure 6-9. An instrumented technique for limit computation, as seen by a teacher

An instrumented technique can be *taught*, but what is taught is not necessarily what students *learn*: the gap between instrumented techniques as taught and as practiced may be important (Appendix 6-2, which shows two students' very different work within the same class and for the same taught instrumented technique).

Describing activity in terms of instrumented action schemes calls for consideration of operational invariants (Box 6-4). A scheme is an observer's construction from the different activity traces of a subject (gestures, anticipations, inferences, etc.). Let us illustrate this, for the same student, in two different environments. He is a student from an experimental class (Trouche 2000), working first for three months with graphic calculators, then for six months with symbolic calculators. The task consists in studying the following question: *"has the given function an infinite limit as x tends to* $+\infty$*?"*.

Graphic calculator environment

If one only described the gestures of the instrumented technique, one would say that the student takes his calculator, 'enters' the function to be studied in the function editor, uses numerical applications to observe the function behavior for large values of the variable, and infers the answers by observing the values taken by f(x) as x takes large values.

If one wants to look at the instrumented action scheme (Figure 6-10), one will search for the operational invariants *guiding* this technique. Searching for them depends on observing the student performing other tasks of the same type and asking him to justify his answers. This student explains he does not use the graphical application, because defining a 'right' window for graphing the function on a large scale is too difficult. Therefore he uses the calculator table of values, and, as far as he can, infers the function behavior. He thus concludes, in the following cases: *"if f(x) is much greater than x, or if the function increases with great speed it is okay. On the other hand, if the function starts to decrease or oscillate then it is no good"*. Consequently one can hypothesize that the student's scheme integrates theorems-in-action of the following type *"if f(x) takes much larger values than x, then the limit of f is infinite"*, *"if the function increases very strongly, then the limit of f is infinite"*, *"if the limit of f is infinite, then f is necessarily increasing"*. From all these properties emerges a concept-in-action of the type: *"f has an infinite limit means that, when x is large, f(x) is very large, increasingly large"*[2].

Symbolic calculator environment

If one describes only instrumented technique, one will say that the student takes his calculator and applies the limit command to the given function.

Concerning the instrumented action scheme (Figure 6-10), there is, compared to the graphic calculator environment, an apparent *simplification*: less effort while manipulating the artifact (the only effort is a syntactic effort of writing a correct command) and less effort of explanation (since the software ensures the correctness of results, any justification of a result, even when required by the teacher, appears less necessary). The instrumentation process leads here to a simplification of the scheme, accompanied by an impoverishment of the operational invariants. To the question: *"what is the meaning of the function having an infinite limit?"* the student, in a graphic calculator environment, gave an answer related to the concept-in-action which we evoked above; four months later, in a symbolic calculator environment, he could not give any definition any more: the function limit did not have any other existence than as a product of the software symbolic application, as a response to a computation command. There was a *vanishing* of the concept. Vanishing does not mean disappearing: the limit conception moves from a *process result*, in a graphic calculator environment, to an *operation result*, in a symbolic calculator environment.

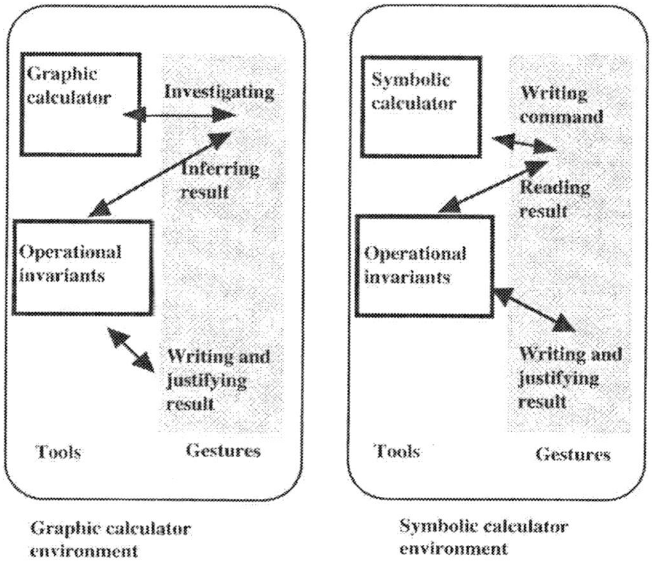

Figure 6-10. Evolution of limit computation action schemes, from a graphic calculator environment to a symbolic one

As we can see for these two instrumented action schemes, there is a dialectic relationship between operational invariants and realized gestures:

- operational invariants *guide* gestures: in the first case, they guide gestures through the investigation process, the inference process and the justifying process. In the second case, they guide gestures of writing a command, reading a result, and (although weak) a process of justifying. In the graphic calculator environment, the mobilization of operational invariants requires an important cognitive effort (one has to evaluate if the x values are large 'enough' and if f(x) is large 'enough'). In the symbolic environment, the cognitive effort is not of the same nature: it is not related to a search process, but only to a control of syntax (here we speak of a particular student; amongst other students, we observed other schemes, Appendix 6-2);
- at the same time, activity, through gestures, *institutes* operational invariants: "From successive approximations, the hand finds the right gesture. The mind registers the results and infers an efficient gesture scheme. Gesture is a synthesis." (Billeter 2002). Operational invariants appear as an abstraction of what is judged an apt gesture. Then, because

operational invariants enable a task to be performed, their field of operationality and validity will naturally spread.

A schema expresses this dialectic between action and conceptualization (Figure 6-11).

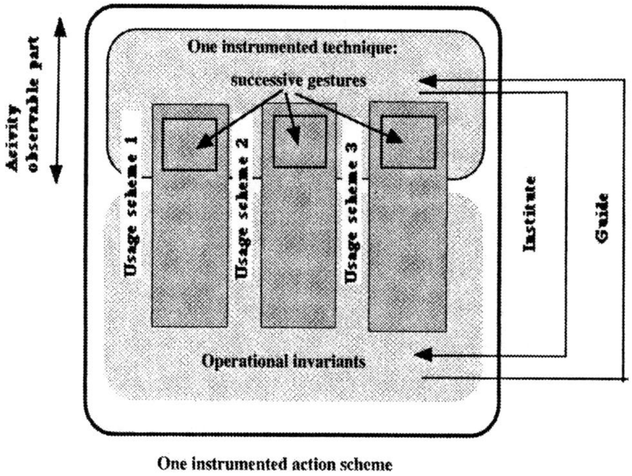

Figure 6-11. Relationships between scheme and technique, gesture and operational invariants

This study of schemes of instrumented action leads to two conclusions:

- the first one relates to the two instruments successively built by the same student: it clearly appears that *extension (or complexification) of an artifact can go with a reduction (or an impoverishment) of the corresponding instrument built by a subject*[3];
- the second one relates to the method of studying instrumented action. The study of instrumented action schemes requires studying, beyond the techniques themselves, their epistemic, heuristic and pragmatic functions (Box 5-1). It requires analysis of the student's activity in more depth: over time, in order to pinpoint regularities, and with regard to the student's discourse, in order to pinpoint the justification offered for gestures. These regularities of activity and justifications of gestures allow hypotheses about operational invariants to be formulated.

Having a good knowledge of calculator constraints and more precise ideas on students' operational invariants may give teachers some means to orient their mathematics lessons:

- choosing situations which help students to master concepts (something which cannot be realized for an isolated concept, but only in the frame of a conceptual field, Box 6-5). This question will be studied in Chapter 9;
- taking into account the artifacts available in the learning environment in order to favor social aspects of schemes. We will see this point in Chapter 8.

Box 6-5.

Operational invariants, concepts and conceptual fields
(Vergnaud 1996)

A *concept* acquires its sense through several situations and phenomena and, thus, has its roots in several categories of operational invariants. Besides, it becomes fully a concept only through articulation of its properties and of its nature, in a mathematical wording where it has the status either of predicate or object.
This idea leads to the definition of a *concept* as a triplet of three sets:
- a set of situations which give sense to the concept;
- a set of operational invariants through which such situations are treated;
- a set of language and symbolic representations which allow the concept to be represented.

A concept cannot be built in isolation. It has to be studied as an element of a larger set which Vergnaud names a *conceptual field* (for example, the limit concept has to be built within the conceptual field of calculus). One can define a conceptual field in a twofold manner:
- as a set of situations needing to be progressively mastered and a closely interconnected range of concepts, procedures and symbolic representations;
- as the set of concepts which ensure a mastering of these situations.

APPENDICES

Appendix 6-1
Some organization constraints for two symbolic calculators

What are, for two different symbolic calculators, the organization constraints related to study of functions?

Computation

TI-92 (Texas Instruments)	Algebra FX 2.0 (Casio)
The same application (available on the keyboard: *HOME*) allows numerical and formal, approximate and exact computation. Within this application, the user can choose the computation mode s/he wants. This mode is indicated at the foot of the screen.	Two different computation applications: first *(NUM)* for approximate computation, second, *CAS*, for formal and exact computation. The computation mode does not remain indicated on the screen when calculating something.

Two different design choices:

- for Casio, approximate mode appears privileged through the first application proposed, *NUM*, only allowing approximate computation;
- for TI, the two modes are placed on the same plane. A keystroke combination (*approximate detour*, § 2.3.1) allows the shift from an exact to an approximate result (from example from 1/3 to 0.33333). Both exact and approximate values can coexist on the same screen.

Between the two cases, a different relationship to numerical approximation is favored.

Combination of commands

TI-92 (Texas Instruments)	Algebra FX 2.0 (Casio)
One can combine approximate and symbolic computation, writing for example the following command: *Approximate value(limit f(x), x, a).*	It is impossible to combine approximate and symbolic computation commands.

These different design choices could have consequences, for example on the conceptualization of the limit notion.

Graphical and numerical analysis of functions

TI-92 (Texas Instruments)	Algebra FX 2.0 (Casio)
GRAPH application (allowing graphically representing function) and *TABLE* application (allowing obtaining table of values) are accessible on the same level on the keyboard.	Opening *GRAPH* application gives access to several menus. One of these menus contains the *TABLE* application. This application thus appears to be included within the *GRAPH* application.

Two different design choices:

- for Casio, graphical representation (compared to table of values) appears privileged;
- for TI, these two types of representation are on the same level;

In these two cases, it is not the same graphical/numerical articulation which is favored.

Appendix 6-2
Two instrumented techniques for computation of limits, in a symbolic calculator environment, for two 12th grade students

(Trouche 2001, p.16)

One can see below the gap between the instrumented technique which is taught (Figure 6-9) and the two students' techniques.

Students had to determine the limit as x tends to +∞ of the function f defined on]0; +∞[by

$f(x) = \dfrac{\sqrt{x} + \cos x}{x + \sin x}$.

NB. The TI-92 symbolic calculator does not "know" this limit (Box 8-4, § 2.1, p.222), but students could use some basic theorems to derive this limit as equal to 0.

Student 1

He first defines the f function for the calculator (Figure 6-12, next page) "*then I could avoid having to write this complex thing several times*". Calculator answer: undef.

"*Oh, these functions sine and cosine often cause trouble when looking for limits, I need to get rid of them*".

On his paper, he bounds sine and cosine as lying between – 1 and +1, and then bounds the f function, for x > 0:

$$\frac{\sqrt{x}-1}{x+1} \le f(x) \le \frac{\sqrt{x}+1}{x-1}$$

He uses his calculator to find the limits of the left and right function: 0.

"*According to the theorem about limits and inequalities, I can say that my function f has also 0 as a limit*".

"*Let us have a look at the graphs of the three functions.* He graphs the three functions: *"the function f is well bounded by the two others in the neighborhood of* +∞*"*.

Then: "*I can also change the variable*". On paper:

$$X = \sqrt{x}, \quad f(X) = \frac{X+1}{X^2+1}$$

"*I can use the theorem about the polynomial functions, or do some factorizations and use the theorems about limits and operations*":

On paper again:

$$\frac{X+1}{X^2+1} = \frac{1+\frac{1}{X}}{X+\frac{1}{X}}$$

End of the work (one hour): paper-and-pencil and calculator approaches articulated, a work in multiple-registers (algebraic and graphical studies), expression and construction of knowledge about limits, a rich limit scheme.

The tool complexity is mastered and contributes to enrich the instrumentation process and to build an efficient instrument for study of function limits.

Student 2

He uses the limit command of the CAS application, applied to the given function.

Calculator answer: undef.

"*Oh, I made a mistake in writing the command!*"

He writes again, same calculator answer.

"Oh, I am useless, I'll have to try again" (writing the function $\dfrac{\sqrt{x}+\cos x}{x+\sin x}$ takes a lot of time). Same calculator answer...

"Oh, I have understood, the calculator doesn't know the f function, I have to define it!".

He defines the function f (Figure 6-12).
Again the command limit, again the answer undef.
New perplexity, and new idea: "*when a limit isn't defined, it is sometimes possible to look at the left, or at the right of the point. So I am going to look at the right of* $+\infty$, *so I will go as far as possible*" (Figure 6-12). Still answer undef.

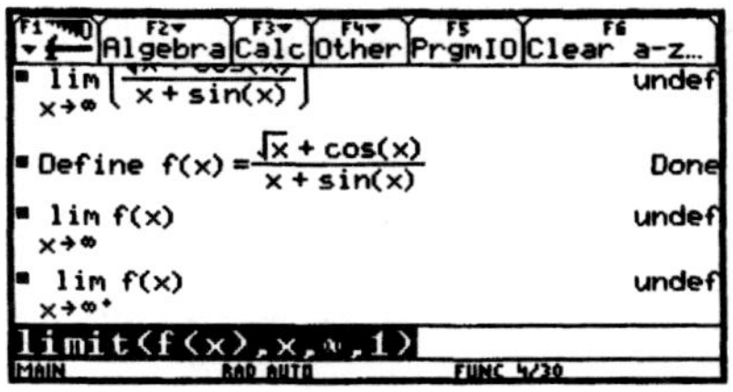

Figure 6-12. Calculator screen copy of student 2

At last, he breaks down the problem into sub-problems, looking for the limits of $\sqrt{x}$ (*"It works, I obtain* $+\infty$ *as a limit!"*) and of sinx and cosx (*"that is the problem: these two functions have no limit, that is the reason why my function f has not limit"*).

End of the work (1hour 30minutes): no paper used, work in a single register (no numerical nor graphical studies), no idea of the function behavior, a quite weak scheme for studying limits.

The complexity of the tool does not contribute to assist the student's activity or build an efficient mathematical instrument.

NOTES

1. "We don't see only forms, but also meanings" (Wittgenstein, in Bouveresse (1995)).
2. This concept-in-action appears close to a *kinematic point of view* on function limit (Box 8-4).
3. This situation is not necessarily linked to a given environment. In the same class, we showed (Trouche 1996) the existence of very different processes for other students: different forms of instrumentalization developed (storage into calculator of the main theorems related to function limits, of specific programs for computation of limits, etc.) and instrumentation becoming richer with the shift from graphic to symbolic calculator environment, through use of a great diversity of applications (Appendix 6-2).

REFERENCES

Artigue M. (1997) Le logiciel DERIVE comme révélateur de phénomènes didactiques liés à l'utilisation d'environnements informatiques pour l'apprentissage, *Educational Studies in Mathematics* 33(2), 133-169.

Balacheff N. (1994) Didactique et intelligence artificielle, *Recherches en Didactique des Mathématiques* 14(1/2), 9-42.

Bernard R., Faure C., Noguès M., Nouazé Y. & Trouche L. (1998) *Pour une prise en compte des calculatrices symboliques en lycée.* Montpellier: IREM, Université Montpellier II.

Billeter J.F. (2002) *Leçons sur Tchouang-Tseu*, Editions Allia.

Bouveresse L. (1995) *Langage, perception et réalité*, Editions Jacqueline Chambon.

Chacon P.R. & Soto-Johnson H. (1998) The Effect of CAI in College Algebra incorporating both Drill and Exploration, *The International Journal of Computer Algebra in Mathematics Education*, 201-216.

Duval R. (2000) Basic Issues for Research in Mathematics Education, in T. Nakahara & M. Koyama (Eds.), *Proceedings of PME 24* (Vol. 1, pp. 55-69). Hiroshima: Hiroshima University.

Dagher A. (1996) Apprentissage dans un environnement informatique: possibilité, nature, transfert des acquis, *Educational Studies in Mathematics* 30, 367-398

Defouad B. (2000) *Etude de genèses instrumentales liées à l'utilisation de calculatrices symboliques en classe de première S,* Doctoral Thesis. Paris: Université Paris VII.

Guin D. & Trouche L. (1999) The complex process of converting tools into mathematical instruments: the case of calculators, *International Journal of Computers for Mathematical Learning* 3, 195-227.

Luengo V. & Balacheff N. (1998) Contraintes informatiques et environnements d'apprentissage de la démonstration en géométrie, *Sciences et Techniques Educatives* 5(1), 15-45.

Noss R. & Hoyles C. (1996) *Windows on Mathematical Meanings, Learning Cultures and Computers* (pp. 153-166). Dordrecht: Kluwer Academic Publishers.

Rabardel P. (1995) *Les hommes et les technologies, approche cognitive des instruments contemporains*. Paris: Armand Colin.

Schwarz B.B. & Dreyfus T. (1995) New actions upon old objects: a new ontological perspective on functions, *Educational Studies in Mathematics* 29(3), 259-291.

Trouche L. (1995) E pur si muove, *Repères-Irem* 20, 16-28.

Trouche L. (1996) *Enseigner les mathématiques en terminale scientifique avec des calculatrices graphiques et formelles*. Montpellier: IREM, Université Montpellier II.

Trouche L. (1997) *A propos de l'enseignement des limites de fonctions dans un "environnement calculatrice", étude des rapports entre processus de conceptualisation et processus d'instrumentation*, Doctoral Thesis. Montpellier: Université Montpellier II.

Trouche L. (2000) La parabole du gaucher et de la casserole à bec verseur: étude des processus d'apprentissage dans un environnement de calculatrices symboliques, *Educational Studies in Mathematics* 41(2), 239-264.

Trouche L. (2001) Description, prescription, à propos de limites de functions, in J.B. Lagrange & D. Lenne (Eds.), *Calcul formel et apprentissage des mathématiques* (pp. 9-26). Paris: INRP.

Vergnaud G. (1996) Au fond de l'apprentissage, la conceptualization, in R. Noirfalise & M.J. Perrin (Eds.), *Actes de l'Ecole d'Eté de Didactique des Mathématiques* (pp. 16-28). Clermont-Ferrand: IREM, Université Clermont-Ferrand II.

Verillon P. & Rabardel P. (1995) Cognition and artifacts: A contribution to the study of thought in relation to instrument activity, *European Journal of Psychology of Education* 9(3), 77-101.

Vygotsky L.S. (1962) *Thought and Language*. Cambridge, MA: MIT Press.

Yerushalmy M. (1997) Reaching the unreachable: technology and the semantics of asymptotes, *International Journal of Computers for Mathematical Learning* 2, 1-25.

Chapter 7

COMPUTER ALGEBRA AS AN INSTRUMENT: EXAMPLES OF ALGEBRAIC SCHEMES

Paul Drijvers and Koeno Gravemeijer
Freudenthal Institute
Utrecht University, The Netherlands
P.Drijvers@fi.uu.nl, K.Gravemeijer@fi.uu.nl

Abstract: In this chapter, we investigate the relationship between computer algebra use and algebraic thinking from the perspective of the instrumental approach to learning mathematics in a technological environment, which was addressed in the previous chapter.

Data comes from a research study on use of computer algebra for developing algebraic insights. Teaching experiments were carried out in ninth- and tenth-grade classes in which the students used symbolic calculators to solve algebraic problems, and in particular for solving parameterized equations and substituting expressions.

We describe in detail instrumented action schemes for solving parameterized equations and substituting expressions. We observe that the approach which students take in their work in the computer algebra environment is closely related to their mental conceptions. The instrumental approach offers ways of making this connection more explicit and better understanding students' difficulties. In particular, we note that students found it difficult to integrate the two schemes into one comprehensive scheme.

We argue that a relationship needs to be established and elaborated between the instrumental approach and other theoretical notions on learning such as the symbolization perspective.

Key words: Algebra, Computer Algebra, Instrumental Genesis, Instrumentation, Mathematics Education, Scheme, Symbolic Calculator, Technology.

1. INTRODUCTION

Computer algebra systems (CAS), both handheld and desktop versions, are finding their way into mathematics classrooms on a regular basis. In the early 1990s, technology became more widespread and software became more user friendly. The interest in the potential of computer algebra for mathematics education grew, and optimism dominated the debate. As has already been indicated in Chapter 1 (§ 1), this technological device was supposed to open new horizons that had previously been inaccessible and to provide opportunities for exploring mathematical situations. In this way, the tool would facilitate students' investigations and discoveries.

This optimism about CAS in the classroom has now taken on additional nuances. The research survey of Lagrange and his colleagues indicates that difficulties arising while using computer algebra for learning mathematics have gained considerable attention (Lagrange & al 2003). For example, Drijvers (2000, 2002) addresses obstacles that students encountered while working in a CAS environment. Heck (2001) notes differences between the algebraic representations found in the computer algebra environment and those encountered in traditional mathematics. He suggests that such differences may provoke conceptual difficulties. Lagrange (2000, Chapter 5) indicates that techniques that are used within the computer algebra environment differ from the traditional paper-and-pencil techniques, which once more may lead to conceptual difficulties.

The integration of computer algebra in mathematics education seems more complicated than one might expect it to be. The idea that we can separate techniques from conceptual understanding and that leaving the first to the technological tool would enable us to concentrate on the latter has been shown to be inadequate and naïve. Rather, we now acknowledge the intertwining of machine techniques and conceptual understanding, which co-evolve simultaneously. In fact, this insight can be considered as the core of the instrumental approach to learning mathematics in a technological environment. According to this approach (addressed in the previous chapter of this book), a process of instrumental genesis involves the development of schemes in which technical and conceptual aspects interact and co-develop.

It goes without saying that, in a computer algebra environment, algebraic schemes are of particular interest. Therefore, one of the goals of our research project *Learning Algebra in a Computer Algebra Environment*[1], which we refer to in this chapter, is to investigate the complex relationship between computer algebra use and algebraic thinking. In this chapter we present some concrete examples of such instrumented action schemes. By doing so, we make concrete the notion of instrumented action schemes, and investigate whether this affords a better understanding of student behavior.

As a second goal of this chapter, we want to reflect on the possible links between the instrumental approach and other theoretical perspectives.

The outline of the chapter is as follows. § 2 revisits the essential elements of the theoretical framework, the instrumental approach. We focus on the notions of schemes and techniques in particular. The design and methodology of the Learning Algebra in a Computer Algebra Environment research project are briefly sketched in § 3. In § 4, 5 and 6 we elaborate the notion of instrumented action schemes by providing concrete examples of algebraic schemes as end products of the learning process. These schemes concern solving parameterized equations (§ 4), substituting expressions (§ 5) and integrating the two into a comprehensive scheme (§ 6). § 7 contains a reflection on the research study and the theoretical framework of instrumentation in particular. It considers possible links with other theoretical perspectives, such as symbolization. Concluding remarks are provided in § 8.

2. THE INSTRUMENTAL APPROACH AS THEORETICAL FRAMEWORK

2.1 Artifact and instrument

The instrumental approach to learning using tools emanates from cognitive ergonomics (Rabardel 1995). The ideas of Vygotsky (1978) on how tools mediate learning can be considered to be the basis of the approach. French mathematics educators (Artigue 1997, Guin & Trouche 1999, Trouche 2000, Lagrange 2000) have applied the instrumental approach to the learning of mathematics using Information Technology (IT). We briefly review the main elements, which have already been developed in Chapter 6 (§ 2).

A central issue in Vygotsky's work is the idea that tools mediate between human activity and the environment. These cultural-historical tools can be material artifacts -- such as calculators or computers -- but also cognitive tools, such as language or algebraic symbols. Rabardel (1995) elaborates on this distinction by stating that a 'bare' *artifact* is not automatically a mediating instrument. The artifact, the material or abstract object, which is given to the user to sustain a certain kind of activity, may be a meaningless object unless the user has used it before or has seen others using it. Only after the user has developed means of using the artifact for a specific purpose, which he considers relevant, does the tool become part of a valuable and useful instrument that mediates the activity and that is build up by the user. The experienced user has developed skills to use the tool in a proficient manner and knows in what circumstances it is useful.

Following Rabardel (1995) and Verillon and Rabardel (1995), we speak of an *instrument* when there is a meaningful relationship between the artifact - or a part of the artifact - and the user for dealing with a certain type of task, in our case mathematical tasks, which the user has the intention to solve. The tool develops into an instrument through a process of appropriation, which allows the tool to mediate the activity. During this process, the user develops mental *schemes* that organize both the problem solving strategy, the concepts and theories that form the basis of the strategy, and the technical means for using the tool. The instrument, therefore, consists not only of the part of the artifact or tool that is involved -- in our case, for example, the algebraic application of a symbolic calculator or CAS -- but can only exist thanks to the accompanying mental schemes of the user -- in our case the student -- who knows how to make efficient use of the tool to achieve the intended type of tasks. The instrument involves both the artifact and the mental schemes developed for a given class of tasks, as symbolized in Figure 7-1. Fig. 6-6 in Chapter 6 provides a different visualization of the same idea.

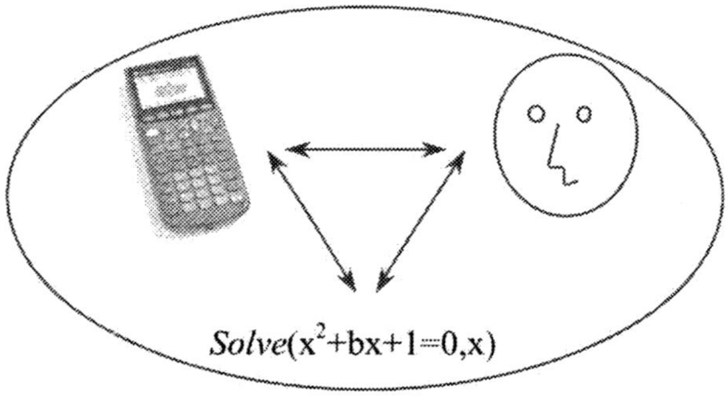

Figure 7-1. The instrument as (part of the) artifact and mental scheme for a class of tasks

It is worth while noticing that the meaning of the word *instrument* here is more subtle than it is in daily life: the artifact develops into an instrument only in combination with the development of mental schemes.

2.2 Instrumented action schemes

As a result of the distinction between artifact and instrument, the 'birth' of an instrument, termed the *instrumental genesis*, is crucial. This instrumental genesis involves the development of mental schemes. But what is such a scheme? How can we identify it and observe its development?

As a first approach to the notion of scheme, we follow Vergnaud, who elaborated on Piaget. Vergnaud defined a scheme as "une organisation invariante de la conduite pour une classe donnée de situations", an invariant organization of activity for a given class of situations (Vergnaud 1987, 1996). A mental scheme has an intention, a goal and it contains different components, such as operational invariants, the often implicit knowledge, which is embedded in a scheme in the form of concepts-in-action or theorems-in-action (Chapter 6, Guin & Trouche 2002, Trouche 2000).

We consider a scheme, therefore, as a stable mental organization, which includes both technical skills and supporting concepts for a way of using the artifact for a given class of tasks. Such schemes are called *utilization schemes* (schèmes d'utilisation in French). Two kinds of utilization schemes can be distinguished (Chapter 6, Guin & Trouche 2002, Rabardel 1995, Trouche 2000). The first category comprises *usage schemes* (schèmes d'usage). A usage scheme is a basic, elementary scheme, which is directly related to the artifact. For example, the moving of a text block while writing in a word processing environment can be done with a cut-and-paste scheme. An experienced user applies this cut-and-paste scheme quickly, accurately and without thinking by means of a sequence of keystrokes and/or mouse clicks. Still, a novice user has to deal with both technical and conceptual aspects, such as knowing the menus or short cuts for the cut and paste commands, but also the frightening fact that the text block that he/she wants to move elsewhere, seems to have disappeared after it has been cut. Accepting the latter requires some insight in the difference between what is on the screen and what is in the memory of the computer.

The usage schemes can serve as building blocks for schemes of the second category, the *instrumented action schemes* (schèmes d'action instrumentée), in which the focus is on carrying out specific kinds of transformations on the objects of activity, which in our case are mathematical objects such as formulas, graphs, etc. Instrumented action schemes are coherent and meaningful mental schemes, and are built up from elementary usage schemes by means of instrumental genesis. This articulation of usage schemes may involve new technical and conceptual aspects, which are integrated in the scheme.

A well-known example of an instrumented action scheme with related conceptual and technical aspects concerns scaling the viewing window of a graphing calculator (Goldenberg 1988). An instrumented action scheme that needs to be developed involves the technical skills of setting the viewing window dimensions, but also the mental skill to imagine the calculator screen as a relatively small window that can be moved over an infinite plane, where the position and the dimensions of the window determine whether we hit the graph. We conjecture that it is the incompleteness of the conceptual part of such a scheme that causes the difficulties that many novice graphing calculator users have with setting appropriate viewing screens. One of the component usage schemes here is a scheme for entering negative numbers with a different minus sign than is used for subtraction. This usage scheme requires insight in the difference between the unary and the binary minus signs.

The difference between usage schemes and instrumented action schemes is not always obvious. Sometimes, it is merely a matter of the level of the user and the level of observation: what at first may seem an instrumented action scheme, may later act as a building block in the genesis of a higher order scheme. For example, further on in this chapter we will describe the integration of two instrumented action schemes for solving equations and substituting expressions into one comprehensive scheme. As an easier example, we refer to the approximation detour (Chapter 6, § 2.3.1) for approximating exact answers by decimal values. When the teaching focuses on this, it could be considered as an instrumented action scheme with difficult conceptual issues such as the difference between real and decimal numbers, and between exactness and accuracy. In a later phase, the approximation detour can be seen as a usage scheme which is integrated into a composite scheme, for example for calculating the zeros of a function which involves the approximation of the outcomes of a *Solve* command.

The examples of the cut-and-paste scheme and the viewing window scheme illustrate that utilization schemes involve an interplay between acting and thinking, and that they integrate machine techniques and mental concepts. In the case of mathematical IT tools, the mental part consists of the mathematical objects involved, and of the mental image of the problem-solving process and the machine actions. The conceptual part of utilization schemes, therefore, includes both mathematical objects and insight into the 'mathematics of the machine'.

The instrumental genesis, in short, concerns the emergence and evolution of utilization schemes, in which technical and conceptual elements co-evolve. Two concluding remarks should be made. First, we point out that the relation between technical and conceptual aspects is a two-dimensional one: on the one hand, the possibilities and constraints of the artifact shape the conceptual development of the user; the conceptions of the user, on the other hand, change the ways in which he or she uses the artifact, and may even lead to changing the artifact or customizing it. These two dimensions are reflected in the difference between instrumentation and instrumentalization (Chapter 6). As a second remark, we should notice that, although the instrumental genesis is often a social process, the utilization schemes are individual. Different students may develop different schemes for the same type of task, or for using a similar command in the technological environment.

In practice, the construction of schemes, the instrumental genesis, is not easy and requires time and effort. Students may construct schemes that are not appropriate, not efficient, or that are based on inadequate conceptions. Examples of difficulties with the instrumental genesis can be found in Drijvers (2000, 2002, 2003b) and Drijvers & Van Herwaarden (2000). In this chapter, we will study such schemes in § 4, 5 and 6.

2.3 Instrumented techniques

As our primary interest in mathematics education concerns the cognitive, conceptual development, the technical side of an instrumented action scheme may seem the less interesting part. However, technical and conceptual aspects interact in an instrumented action scheme. Furthermore, mental schemes cannot be observed or assessed directly, whereas technical actions are visible, observable and can be the subject of teaching. Lagrange, therefore, developed the notion of instrumented techniques, in particular for the case of computer algebra environments, and stresses the fact that techniques change when a technological device is used, but should not be neglected as irrelevant (Lagrange 1999abc, 2000, Chapter 5). He argues that techniques are still important in the computer algebra environment, because they are related to the conceptual aspect by means of instrumented action schemes. Lagrange describes a technique as follows:

> In this link with concepts, the technical work in mathematics is not to be seen just as *skills and procedures*. The technical work in a given topic consists of a set of rules, and methods and, in France, we call such sets *techniques*, as they are less specific and imply less training than *skills* and more reflection than *procedures* (Lagrange 1999c).

In line with this, we see an *instrumented technique* as a set of rules and methods in a technological environment that is used for solving a specific type of problem (Chapter 5). As such, an instrumented technique is the technical side of an instrumented action scheme. The instrumented technique concerns the external, visible and manifest part of the instrumented action scheme, whereas in the instrumented action scheme, the invisible mental and cognitive aspects are stressed. The visibility of instrumented techniques is why they -- rather than the instrumented action schemes, which have a more internal and personal character -- are the gateway to the analysis of instrumental genesis.

3. METHODOLOGY AND DESIGN

One of the research questions of the study *Learning algebra in a computer algebra environment* concerns the instrumental genesis of schemes in a computer algebra environment:

> What is the relation between instrumented techniques in the computer algebra environment and mathematical concepts, as established during the instrumentation process?

In this study we focus on the development of algebraic schemes related to the concept of parameter (Drijvers 2001). As a research design, the paradigm of developmental research -- or design research -- is used, which includes a cyclic design (Gravemeijer 1994, 1998). Each cycle consists of a preliminary phase in which a hypothetical learning trajectory is developed and student activities are designed, a teaching experiment phase in which the learning trajectory is tried in classroom reality, and a retrospective phase in which data is analyzed and 'feed-forward' for the next research cycle is formulated.

The experiments from which the observations in this chapter are drawn took place in the period 2000-2002 in 6 classes. The students were in ninth and tenth grade (14 to 16 year-olds) of the highest, pre-university level of upper secondary education. The grade-nine classes consisted of future students of the exact science stream and of the language or social studies stream grouped together. The tenth-grade classes consisted of students who opted for the exact sciences stream of upper secondary education.

The teaching experiments ran for between three and five weeks with four weekly lessons of forty-five minutes each and dealt with ways in which computer algebra could support the development of insight into the concept of parameter. In this chapter, we focus on two types of algebraic operations: solving systems of equations that may contain parameters, and substituting expressions into equations.

For a better understanding of the episodes described below, we need to explain the approach to algebra adopted in the Netherlands. The introduction of algebra during the first years of secondary education, for students aged 12 to 15, takes place carefully. Much attention is paid to the exploration of realistic situations, to the process of mathematization, and to the development of informal problem-solving strategies. Variables, for example, often have a direct relationship with the context from which they come, and are called price, cost, length, to mention some examples. The translation of the problem situation into mathematics is an important issue, whereas formalization and abstraction are delayed. As a consequence, algebraic skills such as formal manipulations are developed relatively late and to a limited extent. Hence, at the time of the experiments described in this chapter, the students' knowledge of algebraic techniques is limited. For example, the general solution of a quadratic equation has not yet been taught.

The students participating in the experiment had access to the TI-89 symbolic calculator as their computer algebra tool. The rationale for this choice lies in the portability of the machine, its flexibility, and the soundness of the algebra module. The students used the TI-89 both in school and at home. Because they had no previous experience with technology such as graphing calculators, using this type of handheld technology was very new to them.

Data consists of classroom observations (audio and video taped), mini-interviews with students, post-unit interviews with students and teachers, written notebooks, and pre-test and post-test results. In the design phases we identified key assignments. We expressed our expectations on these items beforehand. These expectations were tested by means of mini-interviews with students during the teaching experiments. In the retrospective phase, we investigated patterns in the students' reactions and mini-interviews. In this chapter, we do not provide a full overview of the data and the patterns we found. Instead, we confine ourselves to presenting exemplary observations and mini-interviews, which represent the data in general[2].

4. AN INSTRUMENTED ACTION SCHEME FOR SOLVING PARAMETERIZED EQUATIONS

In this section we discuss the first more elaborated example of an instrumented action scheme. This scheme concerns solving parameterized equations in the computer algebra environment and of course includes technical as well as conceptual aspects. First we discuss two representative classroom observations; then we present the key elements of the scheme as we would like it to emerge from instrumental genesis, together with the main conceptual difficulties that our students encountered in the teaching experiments.

4.1 Prototypical observations

The first observation concerns the task of finding the general coordinates of the intersection points of the graph of $y = ax^2 + bx + c$ with the horizontal axis. The following dialogue shows how this assignment presents two tenth-grade students, Maria and Ada, with problems:

Maria: *This is an extremely difficult question. Find the general coordinates. How can you find coordinates of something if it has no numbers?*

Ada: *[reads the assignment]*

Maria:	*That's impossible, isn't it?*
Ada:	*$ax^2 + bx + c$ ... last year ...*
Maria:	*But how can you know in general where it intersects? That's different for each...*
Ada:	*Intersection points with the x-axis is just filling in zero, that's what they want, isn't it?*
Maria:	*But you can't fill in anything in a formula with a's, b's and c's?*

Even though Maria is, technically speaking, able to solve the parameterized equation with respect to x in the computer algebra environment, a conceptual conflict prevents her from doing so. What Maria expresses in the first sentence of the above protocol is a view of solutions as numerical results instead of something being "different for each" value of the parameters *a*, b and c. Apparently, algebraic expressions cannot be solutions to her.

Applying the TI-89 *Solve* command in this situation requires overcoming this so-called lack-of-closure obstacle (Collis 1975, Küchemann 1981, Tall & Thomas 1991). Maria's conception of solutions of an equation needs to be extended from only numerical results to algebraic expressions. A limited view on solving equations was observed frequently in the teaching experiments of our research study.

Figure 7-2 shows the next assignment in the teaching experiment. In task b the question is similar to the one in the previous assignment, namely to express the coordinates of the zeros in the parameter b.

Below you see a sheaf of graphs of the family $y = x^2 + b \cdot x + 1$.

We pay attention to the extreme values of the parabola.

a. Mark and connect the extreme values. What kind of curve do you seem to get?
b. Express the coordinates of the extreme value of a 'family member' in b.
   Hint: the minimum lies between the zeros, if there are zeros.
c. Find the equation of the curve through all the extrema.

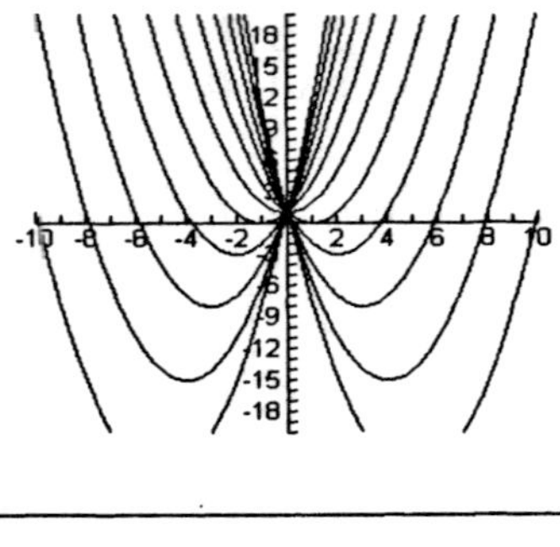

Figure 7-2. The sheaf of graphs assignment

Maria does indeed enter the *Solve* command into the TI-89, but solves the equation with respect to b instead of x:

Maria: *So you do = 0 so to say, and then 'comma b', because you have to solve it with respect to b.*
Observer: *Well, no.*
Maria: *You had to express in b?*

The idea of expressing one variable in terms of others, or, as it is called in the student text, isolating one variable, is apparently not completely clear to Maria. She fails to distinguish the roles of the different literal symbols in an adequate way. Maybe she does not even realize that different values can be substituted for b, each leading to one graph in the sheaf. One can also wonder whether she understands the problem-solving strategy, or is 'blindly' following the suggestions in the assignment text.

Later, however, she solves the equation with respect to the correct unknown (Figure 7-3 upper part). However, copying the result into her notebook leads to an error in the second square root, which indicates that Maria does not understand the structure of the expression in detail (Figure 7-3 lower part).

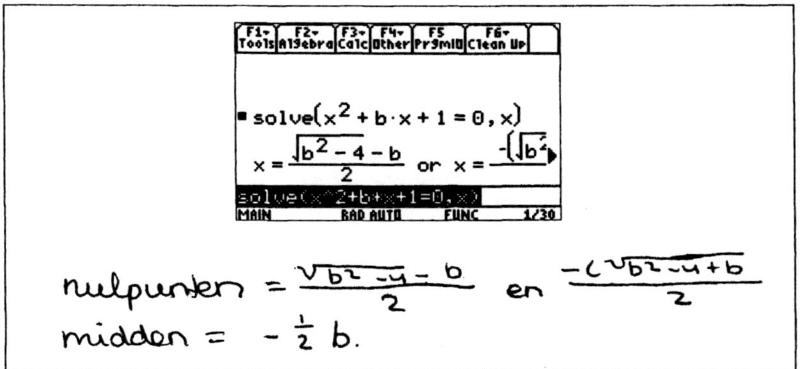

Figure 7-3. The *Solve* command on the TI-89 and the copy in Maria's notebook

4.2 Key elements in an instrumented action scheme for solving parameterized equations

On the basis on an analysis of observations similar to the ones presented in the previous section, we identified some key elements of an adequate instrumented action scheme for solving parameterized equations and the accompanying technique (Figure 7-3), as it might emerge from successful instrumental genesis. Here we ignore aspects that are specific to the assignment presented in the previous section, as this problem situation also encompasses a process of horizontal mathematization. Using Thompson's distinction between relational and calculational reasoning, we limit ourselves here to the latter (Thompson & al 1994). We distinguish the following core elements in an appropriate instrumented action scheme for solving parameterized equations:

1. Knowing that the *Solve* command can be used to express one of the variables in a parameterized equation in other variables.
2. Remembering the TI-89 syntax of the *Solve* command, that is *Solve (equation, unknown)*.
3. Knowing the difference between an expression and an equation.
4. Realizing that an equation is solved *with respect to* an unknown and being able to identify the unknown in the parameterized problem situation.
5. Being able to type in the *Solve* command correctly on the TI-89.
6. Being able to interpret the result, particularly when it is an expression, and to relate it to graphical representations.

In this list some of the elements have a primarily technical character (items 2 and 5) whereas others have a mainly conceptual character. The first item involves an extension of the notion of solving to situations other than the ones already encountered, namely, for expressing one variable in terms of others. Many students have conceptual difficulties in using the *Solve* command to 'isolate' one variable in a parameterized relation that contains other literal symbols. To them 'solve' means finding a numerical solution, whereas an expression is not considered to be a solution. Using computer algebra for this purpose requires the conception of solving to be extended towards a broader view.

The third item concerns being aware of the difference between expressions and equations. Particularly when the equation has the form expression = 0, this distinction may be unclear to students. A lack of awareness of this point may lead to errors in the syntax mentioned in item 2.

The fourth item in the list highlights the -- often implicit -- notion of the 'unknown-to-be-found' (Bills 2001), and is already reflected in the need to specify the unknown in the syntax (item 2). It also addresses the possible shift of roles in the case of a parameterized equation that is to be solved with respect to the parameter. Also, the difference between 'solving with respect to b' and 'express in b' should be clear to the students.

This seemingly technical necessity to add 'comma-letter' to the equation has a mental counterpart: if students are unaware that an equation is always solved *with respect to* a certain variable, they will not understand this requirement and will tend to forget it. However, the fact that the TI-89 requires the letter specification may foster the making explicit of the idea of 'solving with respect to.' Some of the students' explanations (*"you have to get out the x"*, or *"you have to calculate x apart"*) provide evidence for this claim. While using paper-and-pencil, planning to solve an equation and carrying out the solution procedure often take place simultaneously, thus obscuring the implicit choice of the letter with respect to which the equation is solved.

The fifth item concerns the ability to use the technical knowledge of item 2 in combination with the insights referred to in items 3 and 4 to enter a correct command.

The sixth item concerns dealing with the lack-of-closure obstacle and with the ability to perceive algebraic expressions as objects. If students see an algebraic expression as an invitation to a computation process, as a recipe for calculating concrete numerical solutions, they will perceive a lack-of-closure when the result of the *Solve* command is an expression. They are not satisfied with a formula as an object that represents a solution. The ability to see formulas both as processes and as objects is indispensable for algebraic thinking (Gray & Tall 1994, Dubinsky 1991, Sfard 1991). In this case, this *process-object* duality may prevent the students from understanding the solution. In fact, an extension of the conception of a solution, including expressions, is required. Also, the student should expect a result in the form unknown = ….and should be alerted when the result looks different. On the one hand, the process perception of algebraic expressions makes the goal of taking full advantage of the power of computer algebra more difficult to attain. On the other hand, using computer algebra may foster the reification of algebraic expressions.

To summarize this section, we notice that in an appropriate instrumented action scheme for solving parameterized equations technical and conceptual aspects are intertwined. The episodes of student behavior in this section illustrate the differences between the *Solve* command as it is available on the TI-89 calculator and the way in which students perceive the notion of 'solve' from the traditional paper-and-pencil setting. The computer algebra environment seems to foster awareness of elements in the solving process that are often implicit in solving by hand or by head, such as the mathematical equivalence of calculating a numerical solution and expressing one variable in terms of others, the notion that an equation is always to be solved with respect to an unknown, and the object character of algebraic expressions. In this manner, working with computer algebra environments creates opportunities for the teacher to frame these issues as topics for discussion.

Two concluding remarks should be made. First, we stress that reducing an instrumented action scheme to a list of items has the advantage of being concise and concrete, and may help to draw up an inventory of technical and conceptual elements of such a scheme, which may guide teaching and task design. However, it does not do justice to the process of instrumental genesis that leads to its development. In that sense, our representation is limited. Second, we point out that it is not the use of computer algebra *per se* that fosters the instrumental genesis of the instrumented action scheme for solving parameterized equations in the intended way; it is the combination of computer algebra use, task design and educational decisions that together form an appropriate environment for instrumental genesis. For example, reflective discussions on the techniques guided by the teacher may influence collective instrumental genesis in an important way.

5. AN INSTRUMENTED ACTION SCHEME FOR SUBSTITUTING EXPRESSIONS

As a second example of an instrumented action scheme, this section addresses an instrumented action scheme for substituting an algebraic expression into another expression or equation in the computer algebra environment. As was the case for the scheme for solving parameterized equations described in the previous section, technical and conceptual aspects interact. First we discuss an example and some representative classroom observations; then we present the key elements of a scheme as we would like it to emerge from instrumental genesis, together with the main conceptual aspects as we observed them in our teaching experiments.

5.1 Prototypical observations

The observations concern an assignment on the volume of a cylinder, shown in Figure 7-4. Figure 7-5 shows how substitutions, such as the one called for in part b in Figure 7-4, can be carried out on the TI-89. The vertical substitution bar, symbolized by | and read as 'with' or 'wherein', offers a means to substitute numerical values as well as algebraic expressions.

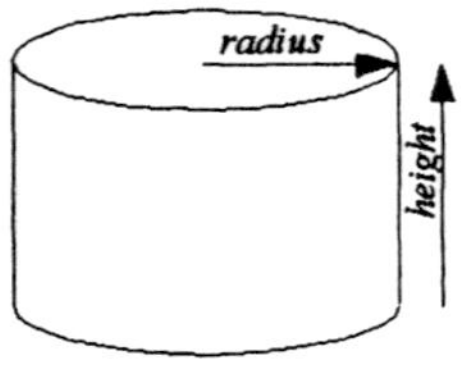

The volume of a cylinder is equal to the area of the base times the height, in short $v = a * h$.
The area of the base equals π times the square of the radius r: $a = \pi * r^2$.

a. Enter these formulas and substitute the formula for the area in that of the volume.
b. If the height of the cylinder equals the diameter of the base, so that h = 2r, the cylinder looks square from the side.
   Express the volume of this 'square cylinder' in terms of the radius.

Figure 7-4. The cylinder assignment

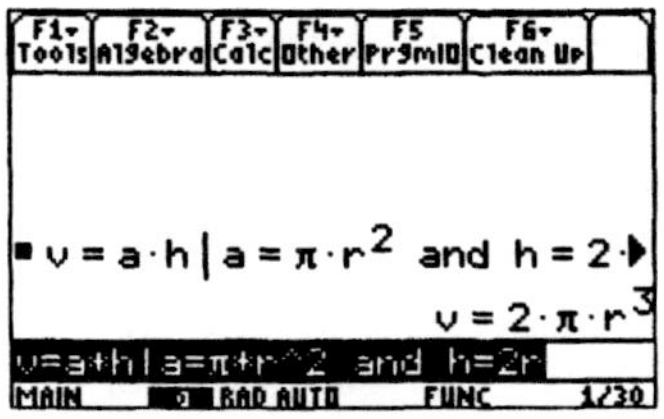

Figure 7-5. Substitution on the TI-89

After a student, Fred, solved part b of Figure 7-4, the teacher asks him to explain the meaning of the substitution bar in a classroom discussion. Fred got help from Cedric:

Teacher: *Now what exactly does that vertical bar mean?*
Fred: *Well, it means that here [after the bar] an explanation is given for that variable.*
Teacher: *You mean the variable is explained, so to say?*
Fred: *Yes. Anyway, that's what I thought and that came out more or less.*
Teacher: *But what do you mean exactly, the variable that is before the bar, or...*
Fred: *Yes the ones before the bar, they are explained here.*
Teacher: *Yeah, that's true, but... Who can explain what exactly the effect of that 'whereby' is?*
Cedric: *Well it, that is in fact a manual for the formula that stands beside [in front] because it says, well, v equals a times h and besides it says yes and there a, that is pi and then r squared.*

Similar explanations are often observed. Students understand the substitution of an expression as *"explaining a variable"*, as *"a guidebook"* or an instruction. Some students come up with vaguer terms such as combining, simplifying or *"making one formula out of it"*. For example, Tony's perception of substitution merely reflects the combining of two formulas:

Observer: *Now what exactly does that vertical bar mean?*
Tony: *It means that the left formula is separated from the right, and that they can be put together.*
Observer: *And what do you mean by putting together?*
Tony: *That if you, that you can make one formula out of the two.*
Observer: *How do you do that, then?*
Tony: *Ehm, then you enter these things [the two formulas] with a bar and then it makes automatically one formula out of it.*

The students who use the vaguer explanations seem to have missed the idea of substitution as visualized in Figure 7-6:

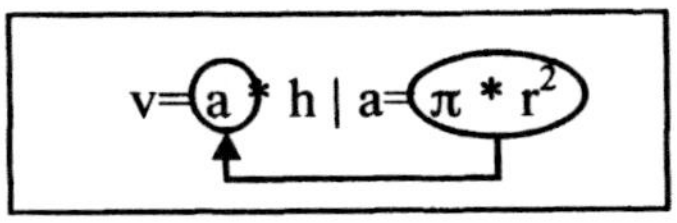

Figure 7-6. Visualization of substitution as 'pasting expressions'

This visualization supports the image of 'cutting an expression' and 'pasting it into' the right position(s) in the expression left of the wherein bar. Our impression is that this visualization contributes to the development of an appropriate conceptual understanding of substitution.

5.2 Key elements in an instrumented action scheme for substituting expressions

On the basis of an analysis of observations similar to the ones presented in the previous section, we identified some key elements of an adequate instrumented action scheme for substituting expressions and the accompanying technique (Figure 7-5), as it might emerge from successful instrumental genesis. Again, we ignore aspects that are specific to the assignment presented in the previous section. We distinguish the following key elements in an appropriate instrumented action scheme for substituting expressions:

1. Imagining the substitution as 'pasting an expression into a variable'.
2. Remembering the TI-89 syntax of the *Substitute* command *expression1* | *variable* = *expression2*, and the meaning of the vertical bar symbol in it.
3. Realizing which expressions play the roles of *expression1* and *expression2*, and considering *expression2* in particular as an object rather than a process.
4. Being able to type in the *Substitute* command correctly on the TI-89.
5. Being able to interpret the result, and particularly to accept the lack of closure when the result is an expression or equation.

As was the case for the instrumented action scheme for solving parameterized equations, some elements of this list have a primarily technical character (items 2 and 4) whereas others have a mainly conceptual character.

The first item involves an extended understanding of substitution. At first, students usually consider substitution to be 'filling in numerical values'. Now, they have to expand this view to the substitution of algebraic expressions, which involves 'pasting expressions'. To put it in a more mathematical way, it involves the understanding of the phenomenon that one of the independent variables in an expression (a in the example of Figure 7-6) in its turn depends on other variables. The use of computer algebra -- in an appropriate educational setting -- on the one hand requires this extension of the notion of substitution and on the other hand may foster it.

In item 3 we recognize once more the process-object issue that we discussed briefly in § 4.2 of this chapter, and the need to perceive an *expression* as an object in order to be able to use the syntax mentioned in item 2. On the one hand, the student must have an object view of *expression2* before he is able to apply the *Substitute* command; on the other hand, the application of the scheme for substituting expressions may foster this object view. Note that it is sufficient for the students to view expressions as objects that can be equated. This does not necessarily imply that they also conceive a function ($a = \pi * r^2$ in the example in the previous section) as an object. The process-object duality is elaborated in more detail in Drijvers (2003a).

In item 5, the student has to deal with the result from the application of the technique mentioned in item 4, and, for example, may notice that all the instances of the *variable* in *expression1* are replaced by *expression2*. If the result still contains the *variable*, the student should be alerted. Once more, the process-object issue and the lack-of-closure obstacle are encountered, as the resulting expression has to be considered as a result, as an expression or equation which symbolizes a relation that often can be used as input for a subsequent process.

To summarize the list of elements, this instrumented action scheme for substituting expressions reveals again an interplay between technical and conceptual aspects. The main conceptual issues in the scheme are the extension of the notion of substitution towards including substitution of expressions, and the object view on the expressions that are substituted. The classroom observations show that the students develop the scheme for substituting expressions relatively easily. The visualization by means of ovals is helpful and the TI-89 notation with the 'wherein bar' is a natural one. The machine technique seems to stimulate the development of the conception of substitution as a replacement mechanism.

We point out that the view of substitution as 'pasting expressions' and the understanding that one of the independent variables in its turn depends on other variables can be seen as the inverse process of the so-called Global Substitution Principle (Wenger 1987). This principle deals with the identification of parts of expressions as units. Maybe the approach of 'placing tiles' over sub-expressions might have been good preparation for the instrumental genesis of an instrumented action scheme for substituting expressions, particularly if the sub-expressions are meaningful in the context.

The substitution activities in the experimental instructional sequence were not limited to substitution as 'pasting expressions'. Students also substituted different parameter values, carried out computations with them and investigated functional relations. However, looking back, we notice that the experimental instructional sequence did not pay attention to a specific kind of substitution: the substitution of relations that were provided in an implicit form. In the next section, where we discuss the integration of a scheme for solving parameterized equations and a scheme for substituting expressions into a more comprehensive scheme, we will encounter the consequences of this omission. As was the case in the previous section, we finish this section by pointing out that our list focuses on the final result of the instrumental genesis and not on the process, and that computer algebra use cannot be separated from the educational context, including task design and teacher guidance.

6. A COMPOSED INSTRUMENTED ACTION SCHEME FOR SOLVING SYSTEMS OF EQUATIONS

In this section we discuss a more comprehensive instrumented action scheme for solving systems of equations. Such a scheme concerns solving systems of equations by means of a strategy which includes application of a scheme for solving equations and a scheme for substituting expressions. Even if the instrumental genesis of the two component schemes has been appropriate, their integration in the composite scheme for solving systems of equations turned out to be far from evident. First we discuss an example and some representative classroom observations; then we present the key elements of a scheme as we would like it to emerge from instrumental genesis, together with the main conceptual aspects as we observed them in our teaching experiments.

6.1 Prototypical observations

Figure 7-7 shows the right-angled triangle assignment which was included in the student texts for the teaching experiments in grade 9. This example is described in more detail in Drijvers & Van Herwaarden (2000).

For this assignment, the textbook proposes an *Isolate-Substitute-Solve* strategy. This technique, abbreviated as ISS, is shown in Figure 7-8. Question a leads to a system of two equations with two unknowns, $x + y = 31$ and $x^2 + y^2 = 25^2$, where x and y stand for the lengths of the two perpendicular edges. First, one of the variables, in this case y, is chosen to be isolated in one of the equations, here $x + y = 31$. The result is then substituted into the other equation using the wherein operator, symbolized by the vertical bar |. In the resulting equation, there is only one variable left (here x) and a second *Solve* command provides the answer, in this case $x = 7$ or $x = 24$. The computation of the corresponding values of the other variable, y is a matter of substituting the values of x in one of the equations. In question b this scheme is repeated for other values, and in c it is generalized for the case with parameters.

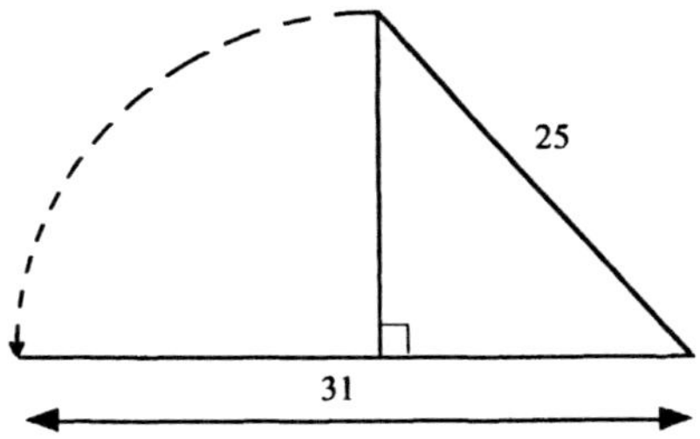

The two right-angled edges of a right-angled triangle together have a length of 31 units. The hypotenuse is 25 units long.

a. How long is each of the right-angled edges?

b. Solve the same problem also in the case where the total length of the two edges is 35 instead of 31.

c. Solve the problem in general, that is without the values of 31 and 25 given.

Figure 7-7. *The right-angled triangle assignment*

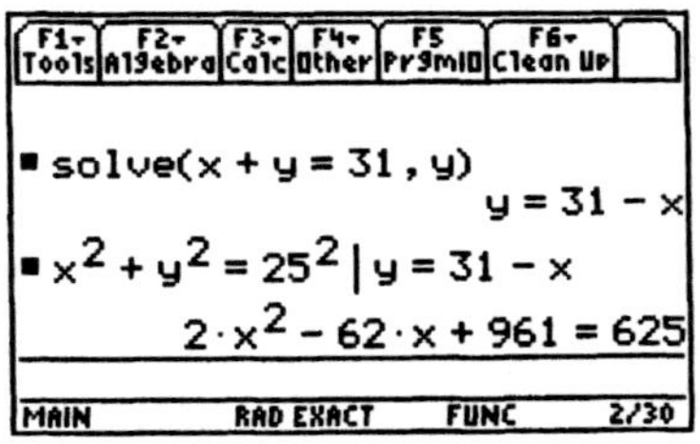

Figure 7-8. The solution scheme ISS on the TI-89

Observations show that some students carry out the first step of this ISS technique -- the isolation -- in their head and substitute the equation $x + y = 31$ immediately as $y = 31 - x$ in the other equation. This, of course, is efficient. However, to students who are not able to do the isolation mentally, the first step of the ISS scheme is not trivial. Many students have conceptual difficulties in using the *Solve* command to isolate one of the two variables. To them 'solve' means finding a solution, whereas $31 - x$ in this context is seen as an expression, not a solution for y. Maybe the concrete geometrical context leads students to expect concrete numerical results more strongly than context-free problems would do. Whatever the reason, the fact that the same *Solve* command is used both for calculating numerical solutions and for isolating a variable requires an extended conception. Evidence presented elsewhere (Drijvers 2003a, 2003b) suggests that even if this conceptual development has been part of the instrumental genesis of the scheme for solving parameterized equations, it is difficult to use this notion in the more complex situation of the comprehensive ISS scheme.

The second step in the proposed ISS scheme, the substitution, presents many difficulties. The frequently observed and persistent error here was that the first step, the isolation, was skipped and the non-isolated equation was substituted directly into the other. The next protocol shows how Rob substitutes the non-isolated form by entering $x^2 + y^2 = 625 \mid x + y = 31$. The CAS does not know which variable in the quadratic equation is to be replaced, so it just returns the quadratic equation. John explains this issue adequately to Rob.

Observer:	*Rob, you wrote down Solve(*$x^2 + y^2 = 625 \mid x + y = 31, y$*). How about that bar?*
John:	*You have to put one apart: you have to put either the x or the y apart.*
Rob:	*You can only work on one letter?*
Observer:	*John, could you explain this to Rob?*
John:	*I think if you have got the vertical bar, you are allowed to explain only one letter, so x =.*
Rob:	*Oh,* $x = 31 - y$.

There are several explanations for this error. First, the two equations in the system play equal, symmetric roles, whereas this symmetry no longer exist in $x^2 + y^2 = 625 \mid x + y = 31$, when the two are treated in a different manner. The students probably see the 'wherein bar', annotated by |, as a symmetric 'with', close to the set theory notation $\{x^2 + y^2 = 25^2 \mid x + y = 31\}$. In fact, it might be better to call this bar the 'cut-and-paste bar' or the 'substitution bar' instead of the 'wherein bar' or the 'with bar'. Second, students may have the idea that while entering the two equations 'the CAS will find a way to combine them'. The analogy with the combined *Solve* command, *Solve*($x^2 + y^2 = 625$ and $x + y = 31, x$) that does work on the TI-89, may play a role here. Third, the immediate substitution obscures the process. Maybe 'lazy evaluation' might help the students: if the result of the substitution $x^2 + y^2 = 625 \mid y = 31 - x$ had been $x^2 + (31 - x)^2 = 625$ instead of $2x^2 - 62x + 961 = 625$, this would provide more insight into the process of substitution as the machine carries it out. However, a mature understanding of substitution would prevent the student from making this error, so once more we see the relation between the syntactic mistake and the conceptual understanding of the process of substituting expressions.

In the final step of the ISS strategy, the solving, a common mistake shows up when the substitution and solving are nested in one combined line. As Rob's input above shows, some students try to combine the substitution and the solution in one line. A correct nested form is:

$$Solve\,(x^2 + y^2 = 25^2 \mid y = 31 - x, x)$$

The problem with this nested form, however, is the choice of the unknown at the end. With respect to which letter do we need to solve? Several times, students chose the wrong letter, the y instead of the x, after the comma. In such instances, the substitution is performed. However, because the y does not appear any longer in the resulting equation, it cannot be solved with respect to y; therefore, it is returned in an equivalent form, which is not always easy to understand (Figure 7-9). The original problem is symmetric in x and y, but in the problem solving process x and y acquire asymmetric roles.

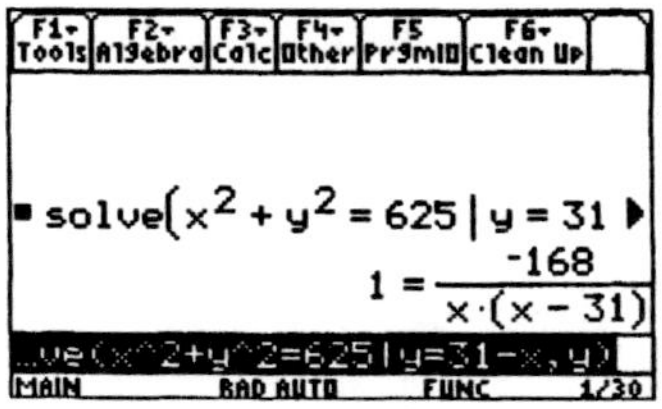

Figure 7-9. Solving for the wrong unknown in the nested ISS technique

To avoid this error, students need to mentally split the nested command into two sub-processes. They need to realize that if the substitution part starts with 'y =', the resulting equation will no longer contain y; therefore, it must be solved with respect to x. This problem-solving behavior shows that students' insight into the substitution process is not mature enough to handle this more complicated situation satisfactorily. The power of computer algebra to integrate two steps requires an increased awareness of what is happening. One of the students avoided this mistake by symbolizing the two sub-processes in the command with an extra pair of brackets:

$$\mathit{Solve}\ ((x^2 + y^2 = 25^2 \mid y = 31 - x), x).$$

One of the teachers in the teaching experiments privileged the nested form of the ISS technique. At the first instance of the problem situation she immediately demonstrated the nested form, which led to many more errors in this class compared to the other class. Observations such as this support the findings of Doerr & Zangor (2000), who argue that the teacher influences the classroom culture and the collective orchestration (used in a broader sense here than in Chapter 9), and the results of Kendal & Stacey (1999, 2001), who state that teachers privilege specific techniques and schemes, and de-privilege others.

6.2 Key elements in an instrumented action scheme for solving systems of equations

On the basis on an analysis of observations similar to the ones presented in the previous section, we identified some key elements of an adequate instrumented action scheme for solving systems of equations and the accompanying technique (Figure 7-9), as it might emerge from successful instrumental genesis. Again, we ignore aspects that are specific to the assignment presented in the previous section. We distinguish the following key elements in an appropriate instrumented action scheme for solving systems of equations:

1. Knowing that the ISS strategy is a way to solve the problem, and being able to keep track of the global problem-solving strategy in particular.
2. Being able to apply the technique for solving parameterized equations for the isolation of one of the variables in one of the equations.
3. Being able to apply the technique for substituting expressions for substituting the result from the previous step into the other equation.
4. Being able to apply the technique for solving equations once more for calculating the solution.
5. Being able to interpret the result, and particularly to accept the lack of closure when the solution is an expression.

This list does suggest an order, as the sequence of steps 2 to 4 cannot be changed. We notice that steps 2, 3 and 4 are instrumented action schemes, which were described in the previous sections. They include technical and conceptual aspects and are embedded here in a composite ISS scheme.

In item 1 we notice that students may have difficulties with keeping track of the overall strategy. For example, students may try to substitute the isolated form into the equation from which it has been derived (Figure 7-10). Of course, the message 'true' causes some confusion, and often students do not understand the logic behind it. Once more, we conjecture that lazy evaluation, in this case leading to x + (31 − x) = 31, would facilitate understanding.

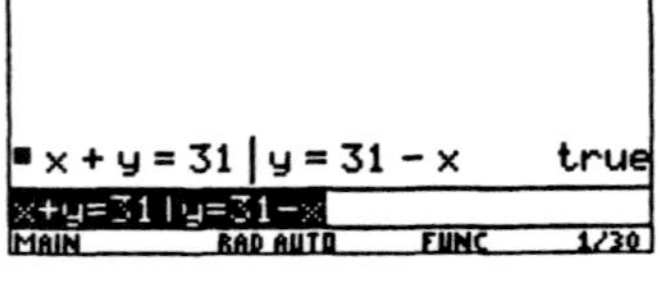

Figure 7-10. 'Circular substitution' in the ISS problem-solving strategy

As an example of such circular substitution, the following dialogue shows that Donald did not know how to solve the system of equations $x*y = 540$, $x^2 + y^2 = 39^2$:

Observer: *What would you do now?*
Donald: *Ehm, well, I calculated that x is 540 divided by y.*
Observer: *Exactly.*
Donald: *Well, and if ... you then can replace this [x in x*y = 540] by that [540/y].*
Observer: *Well, yes, but that is not sensible, because x is 540 divided by y but you shouldn't fill that in the equation where you got that from, where you derived it from.*
Donald: *Oh.*
Observer: *So you'd better substitute it into the other equation.*
Donald: *In that one [the quadratic equation].*
Observer: *Right. So if you fill in this [540/y] here [x in $x^2+y^2=39^2$].*
Donald: *... and that one below [y = 540/x] there [the y in the quadratic equation].*

The last line of the above dialogue reveals a second strategic error of 'double work': by means of two 'isolations', both x and y are isolated from $x*y = 540$. Then x in $x^2 + y^2 = 39^2$ is replaced by 540/y and simultaneously y is replaced by 540/x. That clearly is double work with no progress towards the result. Both the circular approach and the double-work error show a lack of overview of the solution strategy and of the composite scheme as a whole.

In items 2 and 3 of the list, we have already noted the error of forgetting the isolation and substituting a non-isolated form directly. Apparently, embedding a substitute scheme into a more comprehensive ISS scheme, where it is tangled up between two *Solve* commands, causes this error to occur frequently.

The most frequent error in the solution step 4 of the scheme was to solve the equation with respect to the wrong unknown, which does not appear in the equation any longer. This was observed particularly when the students used the nested form of an ISS scheme, in which the substitution and the solving are combined in one command. The compactness of the nested form increases the complexity of the technique.

In step 5 we encounter the process-object duality again in a way similar to the simple instrumented action schemes which constitute the ISS scheme.

We cannot end the discussion of an ISS scheme without pointing out the variations that may exist, and that were indeed observed, in the classrooms. For example, some students prefer to solve both equations first with respect to the same unknown, and then to equate the two right-hand sides. This variation, which we called *Isolate-Isolate-Equal-Solve*, comes down to the following in the example of Figure 7-7:

Isolate twice with respect to y: $y = \sqrt{25^2 - x^2}$ and $y = 31 - x$

Equate both right-hand sides: $\sqrt{25^2 - x^2} = 31 - x$

Solve with respect to x: x = 24 or x = 7.

To summarize this section, we see that a comprehensive scheme for solving systems of equations is more than just the sum of its simple components. The complexity increases, which may lead to errors in the component schemes, even if these errors seemed to have been overcome in the instrumental genesis of these component schemes. Also, keeping track of the overall problem solving strategy may be difficult. For example, the extension of the conception of solving towards isolation of the unknown or 'expressing one variable in terms of the others' is probably more difficult if the outcome of the isolation is an intermediate result than if it is the final outcome of an exercise. This illustrates the interaction between machine technique and conceptual development in the instrumented action scheme: the fact that the same command is used for isolating an unknown as for calculating the final outcome indicates that these operations -- quite different in the eyes of the student -- are identical in the world of computer algebra and are mathematically equivalent. For substitution, a similar remark can be made. Students regularly try to substitute an isolated form into the equation from which it originated, or they try to substitute a non-isolated form. These errors reveal a limited awareness of the problem solving strategy or a limited conception of substitution. Again, we perceive these errors as instrumentation matters. If students have an appropriate mental image of substitution, they should be able to perform the substitutions in the computer algebra environment correctly. Classroom discussion of these instrumentation problems can improve students' understanding of substitution and strengthen the machine skills needed to solve problems like those shown above. As was the case for the simple instrumented action schemes, this shows that the instrumental genesis as well as the resulting schemes may depend largely on the educational setting, the tasks and the role of the teacher.

7. REFLECTION

In this section we first reflect on the findings in our study. Then we reflect on the merits of the instrumental approach and on its possible links with other theoretical frameworks.

7.1 Reflections on the findings of the study

In the previous sections we observed a close relationship between machine techniques and conceptual understanding within the instrumented action schemes. Students can only understand the logic of a technical procedure from a conceptual background. Seemingly technical difficulties often have a conceptual background, and the relation between technical and conceptual aspects makes the instrumental genesis a complex process. We illustrated this by drawing up lists of key elements in instrumented action schemes. Although such lists tend to have a prescriptive and rigid character, and ignore the process of individual instrumental genesis, they were found to be helpful for making explicit the relation between technical and conceptual components, and for indicating possible 'end products' of the instrumental genesis.

The students in our study encountered many obstacles during this instrumentation process, which can be explained in terms of the technique-concept interaction. We argue that such obstacles offer opportunities for learning, which can be capitalized on by reflecting on their conceptual aspects and the relation with the corresponding paper-and-pencil technique. Of course, the teacher plays an important role in turning obstacles into opportunities.

The examples of the instrumented action scheme for solving parameterized equations and substituting expressions show how the mixture of technical and conceptual aspects may lead to difficulties that are not easy for the students to overcome and often demand conceptual development. The development of the instrumented action scheme for solving systems of equations indicates that the integration of elementary schemes into a composite scheme requires mastering of the component schemes at a high level.

Two comments should be made on these conclusions. First, one can ask whether computer algebra is an appropriate tool for algebra education if the instrumental genesis requires so much effort. Indeed, the CAS environment has a top-down character in the sense that 'everything is already there'. No symbolizations have to be developed and the computer algebra environment is not flexible with respect to notational or syntactic differences. The CAS does not provide insights into the way in which it obtains its results. As a consequence, the students may suffer from a perceived lack of transparency and congruence, so that the CAS is seen as a black box. Although the flexibility of any IT tool is limited, this issue is striking for CAS use at this level -- ninth and tenth grade -- of education. Maybe other technological tools, for example environments that provide intermediate results and show 'black box behavior' to a lesser extent, might foster algebraic insight better for students at this educational level.

The second comment concerns the embedding of CAS use in the educational setting. As indicated by Doerr & Zangor (2000), the results of the integration of technology into mathematics education depend on the coordination of many factors, such as the learning goals, the available instructional tasks, the didactical contract (Box 8-6), the socio-mathematical norms in the classroom, the guidance provided by the teacher and the previous algebra education, to mention some. For the integration of CAS in education, we recommend realistic problem situations as points of departure, which may foster the development from informal meanings and strategies to more formal methods. In this way the contexts, which first function at a referential level, may acquire a general meaning. We conjecture that the instrumentation difficulties that we reported in the § 4, 5 and 6 might appear in different ways in such an educational approach, depending on changes in some of the factors involved; we do believe, however, that these difficulties will be encountered in one way or another during the instrumental genesis.

7.2 Reflection on the instrumental approach

§ 4, 5 and 6 illustrate how the instrumental approach can be used as a framework for observing, interpreting and understanding students' behavior in the computer algebra environment. We acknowledge that we used the approach in this chapter only to a limited extent, namely for focusing on the individual algebraic schemes as end products rather than taking the process perspective. We neglected the embedding of these schemes in the educational process as well as the process of instrumental genesis. The roles of the teacher and of the classroom community were hardly addressed.

In the present section, we reflect on the role of this theoretical framework, and we discuss possible links with other theoretical approaches. The latter issue is important, as we agree with Hoyles and Noss who consider the connection between research on technology use and trends in research on the learning of mathematics in general as highly relevant:

> Our aim therefore, is to bring the field of research with and on computationally-based technologies in mathematical learning closer to the broader field of mathematics education research. We take it as axiomatic that each has much to learn from the other; but we are fully aware of just how insulated the work with digital technology has been (Hoyles & Noss 2003).

What does the instrumental approach to learning mathematics using technological tools allow for? In our opinion, it stresses the subtle relationship between machine technique and mathematical insight, and provides a conceptual framework for investigating the development of schemes, in which both aspects are included. Important elements of this framework are the notion of instrument, the process of instrumental genesis and the distinction between different kinds of schemes. These elements are, in our opinion, helpful for designing student activities, for observing the interaction between students and the computer algebra environment, for interpreting it and for understanding what works out well and what does not.

The seemingly technical obstacles that students encounter while working in the computer algebra environment often have conceptual components, and the instrumental approach helps to be conscious of this and to turn such obstacles into opportunities for learning.

After noticing the power of the instrumental approach, we now reflect on the directions for future development, and on issues which might need further investigation. Below, we address the questions of task design for provoking instrumental genesis, of the scope of the instrumental approach, of the accessibility of the theory, and of the need for articulating the instrumental approach with other theoretical perspectives such as theories on symbolizing, cultural-historical activity theory, and the socio-constructivist framework.

The first issue concerns the question of task design. The instrumental approach investigates the intertwining of technical and conceptual aspects within instrumented action schemes, and the process of developing such schemes, the instrumental genesis. Therefore, important questions for classroom practice are: What do we know about the design of instructional activities which foster instrumental genesis? What characteristics can we identify in tasks that enhance a productive instrumental genesis? The articulation of the instrumental approach and the notion of didactical engineering is addressed in Chapter 9.

The second issue is the scope of the instrumental approach. We notice that the instrumental approach to learning mathematics using technological tools is developed and applied particularly in the context of using computer algebra. Although it has also been applied to the use of dynamic geometry software (Laborde 2003, Mariotti 2002), it is not yet completely clear how general its applicability is. What would the instrumental approach offer us for the educational use of other types of software environments, such as Java applets or statistical software? How does the instrumental approach relate to different types of tool use? For example, while studying the integration of the graphing calculator, Doerr & Zangor (2000) distinguish the computational tool, the transformational tool, the data collection and analysis tool, the visualizing tool and the checking tool. What does instrumental genesis mean for each of these types of tool use? These questions deserve further investigation.

Let us elaborate on this. Computer algebra has specific features, such as its rigidity concerning input syntax and, in the case of the TI-89, which we used, its direct evaluation of input without possibilities for providing intermediate results. For educational purpose, one could use other, more open software environments for developing the insights that proficient computer algebra use requires, such as Java applets. For example, one might want to anticipate the object view on formulas and expressions, which is privileged by the computer algebra environment, more at a functional level using other types of technological tools. As an aside, we can draw the analogy with data analysis. We conjecture that the use of data analysis software packages leads to similar instrumentation problems as CAS use does. To choose an adequate statistical operation or representation, the student has to be able to anticipate how these might help in answering the problem at hand. This, in turn, requires a deeper statistical understanding, which encompasses, for instance, a well-developed conception of a distribution of data points, within which a distribution is conceived as an object-like entity. Research shows that pedagogical software tools can be used to foster the development of such a notion of distribution in parallel with the corresponding graphical representations and statistical tools to prepare students for the use of ready-made statistical toolboxes (Cobb, McClain & Gravemeijer 2003, Bakker & Gravemeijer in press).

As a third comment on the instrumental approach, we point out that the accessibility to novice researchers might be improved by considering our terminology. We have already mentioned the quite specific meaning of the word instrument. As a second example, we mention the difference between instrumentation and instrumentalization, which is far from self-evident from the wording. In line with the terminology of Hoyles & Noss (2003), would it not be more clear to just speak about the two-sided relationship between tool and learner as a process in which the tool in a manner of speaking *shapes* the thinking of the learner, but also *is shaped by* his thinking? In order to disseminate it further, the terminology of the instrumental approach might be reconsidered.

The fourth comment concerns the articulation of the instrumental approach with other theoretical perspectives. For example, many current research studies on mathematics education focus on semiotics, symbolizing, modeling and tool use (e.g. Cobb & al 1997, Gravemeijer 1999, Gravemeijer & al 2000, Meira 1995, Nemirovsky 1994, Roth & Tobin 1997). These approaches stress the dialectic relation between symbolizing and development of meaning in increasing levels of formalism. As our example of data analysis tools illustrate, this relation may be problematic if in the tool 'everything is already there'. Computer algebra offers limited possibilities for the student to develop his own informal symbolizations. The instrumental approach stresses the development of schemes in a fixed technological environment, but does not seem to pay much attention to the 'symbolic genesis', the development of symbols and their related meanings. However, instrumental genesis includes a signification process of giving meaning to algebraic objects and procedures. Therefore, we suggest a study, which addresses the compatibility or incompatibility of the instrumental approach and the perspective of symbolizing, both within and apart from the technological environment.

To what other theoretical perspectives do we think the instrumental approach might be connected? Utilization schemes have both an individual and a social dimension. As utilization schemes are mental schemes, they are essentially individual, and instrumental genesis is an individual process. In the meantime, instrumental genesis takes place in a social context. As Rabardel & Samurcay (2001) phrase it: "... their emergence results from a collective process that both the users and designers of artifact contribute to". The articulation of these two dimensions, and in particular of the instrumental approach with cultural-historical activity theory (Engeström & al 1999) and with the notion of the *community of practice* (Wenger 1998), deserves much attention. The elaboration of the concept of orchestration in Chapter 8 can be seen as a good step in this direction.

In connection with this we refer to the interpretative framework that is developed by Yackel & Cobb (1996). They show how an individual, psychological, perspective can be coordinated with a social perspective that takes the group as a unit of analysis. This coordination is elaborated in an analysis of the reflexive relation between social norms, socio-mathematical norms and mathematical practices on the one hand, and individual beliefs and insights on the other. On the basis of this interpretative framework, questions with respect to the relations between classroom norms, the way technology is used by the students, and the role of the teacher in this, may be investigated. Preliminary results of studies that take into account the role of the teacher (Doerr & Zangor 2000, Kendal & Stacey 1999, 2001) suggest that this perspective might add value.

To summarize this reflection, we note that, although we applied the instrumental approach only in a limited way, it provided help in observing, interpreting and understanding the student-machine interactions. However, in order to fully investigate its relevance, we recommend further research on applying the approach to other technological environments and to relate it to other relevant theoretical perspectives to research on the learning of mathematics in general.

8. CONCLUSION

In this chapter we argue that the understanding of algebraic concepts and computer algebra techniques are closely related to one another. Difficulties while carrying out computer algebra techniques could often be linked to limitations in the students' algebraic insight. The development of mental conceptions, on the other hand, could be fostered by the machine techniques.

The instrumental approach to learning mathematics using technological tools made explicit this close relation. As such, it is an adequate framework to better understand students' difficulties. In fact, an error that occurs while using a machine may reveal a lack of congruence between machine technique and mathematical conception, or may indicate limitations in the conceptualization of the mathematics involved. The concept of instrumented action schemes, which involve both machine techniques and mental conceptions, is a powerful one for making concrete the interactivity between machine technique and conceptual understanding.

By means of providing examples of instrumented action schemes -- though presented as end products rather than as processes of instrumental genesis -- we show how the instrumental approach and the notion of schemes in particular may help the observer -- whether he is a researcher or a teacher -- to interpret and understand what is happening while students work in the computer algebra environment. The lists of key elements of instrumented action schemes are not intended to suggest that such schemes have a rigid and universal character. They do provide examples of how the decomposition of elements within the schemes can help the observer to analyze the complexity of the students' work in the technological environment.

Our conclusion, therefore, is that the instrumental approach provides a fruitful framework in the research study that we report on. In the meantime, however, we have identified some limitations. So far, the instrumental approach seems to suffer from a somewhat isolated position, and a too close link to research on the integration of computer algebra into mathematics teaching. Therefore, we recommend investigating the usefulness of the approach in studies that focus on other technological devices, and linking the instrumentation approach to other theoretical perspectives from research on mathematics education.

Paul DRIJVERS (P.Drijvers@fi.uu.nl) is a researcher in mathematics education at the Freudenthal Institute (Utrecht University, the Netherlands). His main research interest is the integration of technologies such as graphing calculators and computer algebra systems in mathematics education at upper secondary level. Besides his research activities, Paul is involved in in-service teacher training.

Koeno GRAVEMEIJER (K.Gravemeijer@fi.uu.nl) is a research coordinator at the Freudenthal Institute and a professor in the faculty of Social Sciences (both at Utrecht University, the Netherlands). His interests concern the role of symbolizing and modeling in mathematics education, and research approaches that integrate instructional design and research.

NOTES

1. Granted by The Netherlands Organization for Scientific Research (NWO) under number 575-36-003E and described in detail in the dissertation of the first author (Drijvers, 2003b).
2. For a more detailed description of the data see (Drijvers, 2003b) or (www.fi.uu.nl/~pauld/dissertation).

REFERENCES

Artigue M. (1997) Rapports entre dimensions technique et conceptuelle dans l'activité mathématique avec des systèmes de mathématiques symboliques [Relations between technical and conceptual dimensions in mathematical activity with systems of symbolic mathematics], *Actes de l'université d'été 1996* (pp. 19-40). Rennes: IREM, Université de Rennes.

Bakker A. & Gravemeijer K.P.E. (in press) Learning to Reason about Distribution, in D. Ben-Zvi & J. Garfield (Eds.), *The Challenge of Developing Statistical Literacy, Reasoning, and Thinking*. Dordrecht: Kluwer Academic Publishers.

Bills L. (2001) Shifts in the meanings of literal symbols, in M. Van den Heuvel-Panhuizen (Ed.), *Proceedings of PME 25* (Vol. 2, pp. 161-168). Utrecht: Freudenthal Institute.

Cobb P., Gravemeijer K.P.E., Yackel E., McClain K. & Whitenack J. (1997) Mathematizing and symbolising: The emergence of chains of signification in one first-grade classroom, in D. Kirshner & J.A. Whitson (Eds.), *Situated cognition theory: Social, semiotic, and neurological perspectives* (pp. 151-233). Hillsdale, NJ: Lawrence Erlbaum Associates.

Cobb P., McClain K. & Gravemeijer K. (2003) Learning About Statistical Covariation, *Cognition and Instruction* 21(1), 1-78.

Collis K.F. (1975) *The development of formal reasoning*. Newcastle, NSW: University of Newcastle.

Doerr H.M. & Zangor R. (2000) Creating meaning for and with the graphing calculator, *Educational Studies in Mathematics* 41, 143-163.

Drijvers P. (2000) Students encountering obstacles using a CAS, *International Journal of Computers for Mathematical Learning* 5, 189-209.

Drijvers P. (2001) The concept of parameter in a computer algebra environment, in M. Van den Heuvel-Panhuizen (Ed.), *Proceedings of PME 25* (Vol. 2, pp. 385-392). Utrecht: Freudenthal Institute.

Drijvers P. (2002) Learning mathematics in a computer algebra environment: obstacles are opportunities, *Zentralblatt für Didaktik der Mathematik* 34(5), 221-228.

Drijvers P. (2003a) Algebra on a screen, on paper and in the mind, in J. Fey, A. Cuoco, C. Kieran, L. McMullin & R.M. Zbiek (Eds.), *Computer Algebra Systems in secondary school mathematics education* (pp. 241-267). Reston, VA: National Council of Teachers of Mathematics.

Drijvers P.H.M. (2003b) *Learning algebra in a computer algebra environment. Design research on the understanding of the concept of parameter*, Dissertation. Utrecht: CD-b press.

Drijvers P. & Van Herwaarden O. (2000) Instrumentation of ICT-tools: The case of algebra in a computer algebra environment, *International Journal of Computer Algebra in Mathematics Education* 7, 255-275.

Dubinsky E. (1991) Reflective abstraction in advanced mathematical thinking, in D. Tall (Ed.), *Advanced mathematical thinking* (pp. 95-123). Dordrecht: Kluwer Academic Publishers.

Engeström Y., Miettinen R. & Raija-Leena P. (1999) *Perspectives on activity theory*. New York: Cambridge University Press.

Goldenberg E.P. (1988) Mathematics, metaphors, and human factors: Mathematical, technical and pedagogical challenges in the educational use of graphical representation of functions, *Journal of Mathematical Behavior* 7, 135-173.

Gravemeijer K.P.E. (1994) *Developing realistic mathematics education*. Utrecht: CD-b Press.

Gravemeijer K.P.E. (1998) Developmental research as a research method, in J. Kilpatrick & A. Sierpinska (Eds.), *Mathematics education as a research domain: A search for identity, An ICMI Study* (Vol. 2, pp. 277-295). Dordrecht: Kluwer Academic Publishers.

Gravemeijer K. (1999) How emergent models may foster the constitution of formal mathematics, *Mathematical Thinking and Learning* 1, 155-177.

Gravemeijer K.P.E., Cobb P., Bowers J. & Whitenack J. (2000) Symbolising, modeling and instructional design, in P. Cobb, E. Yackel & K. McClain (Eds.), *Symbolising and communicating in mathematics classrooms: Perspectives on discourse, tools, and instructional design* (pp. 225-273). Hillsdale, NJ: Lawrence Erlbaum Associates.

Gray E.M. & Tall D.O. (1994) Duality, ambiguity, and flexibility: A 'proceptual' view of simple arithmetic, *Journal for Research in Mathematics Education* 25, 116-140.

Guin D. & Trouche L. (1999) The complex process of converting tools into mathematical instruments: The case of calculators, *International Journal of Computers for Mathematical Learning* 3, 195-227.

Guin D. & Trouche L. (2002) Mastering by the teacher of the instrumental genesis in CAS environments: Necessity of instrumental orchestrations, *Zentralblatt für Didaktik der Mathematik* 34(5), 204-211.

Heck A. (2001) Variables in computer algebra, mathematics and science, *International Journal of Computer Algebra in Mathematics Education* 8, 195-221.

Hoyles C. & Noss R. (2003) What can digital technologies take from and bring to research in mathematics education? In: A.J. Bishop, M.A. Clements, C. Keitel, J. Kilpatrick & F. Leung (Eds.), *Second International Handbook of Mathematics Education* (Vol. 1, pp. 323-349). Dordrecht: Kluwer Academic Publishers.

Kendal M. & Stacey K. (1999) Varieties of teacher privileging for teaching calculus with computer algebra systems, *International Journal of Computer Algebra in Mathematics Education* 6, 233-247.

Kendal M. & Stacey K. (2001) The impact of teacher privileging on learning differentiation with technology, *The International Journal of Computers for Mathematical Learning* 6, 143-165.

Küchemann D.E. (1981) Algebra, in K.M. Hart (Ed.), *Children's understanding of mathematics: 11-16*. London: John Murray, 102-119.

Laborde C. (2003) *The design of curriculum with technology: lessons from projects based on dynamic geometry environments*, paper presented at the CAME 2003 conference, Reims, France. Retrieved on 4 February 2004 (http://ltsn.mathstore.ac.uk/came/events/reims/).

Lagrange J.B. (1999a) Learning pre-calculus with complex calculators: mediation and instrumental genesis, in O. Zaslavsky (Ed.), *Proceedings of PME 23* (Vol. 3, pp. 193-200). Haifa: Technion.

Lagrange J.B. (1999b) Complex calculators in the classroom: theoretical and practical reflections on teaching pre-calculus, *International Journal of Computers for Mathematical Learning* 4, 51-81.

Lagrange J.B. (1999c) Techniques and concepts in pre-calculus using CAS: a two-year classroom experiment with the TI-92, *The International Journal of Computer Algebra in Mathematics Education* 6, 143-165.

Lagrange J.B. (2000) L'intégration d'instruments informatiques dans l'enseignement: une approche par les techniques [The integration of computer instruments in education: an approach by means of techniques], *Educational Studies in Mathematics* 43, 1-30.

Lagrange J.B., Artigue M., Laborde C. & Trouche L. (2003) Technology and Mathematics Education: a Multidimensional Study of the Evolution of Research and Innovation, in A.J. Bishop, M.A. Clements, C. Keitel, J. Kilpatrick & F.K.S. Leung (Eds.), *Second*

International Handbook of Mathematics Education (Vol. 1, pp. 239-271). Dordrecht: Kluwer Academic Publishers.

Mariotti M.A. (2002) The influence of technological advances on students' mathematics learning, in L.D. English (Ed.), *Handbook of international research in mathematics education* (pp. 695-723). Mahwah, NJ: Lawrence Erlbaum Associates.

Meira L. (1995) The microevolution of mathematical representations in children's activity, *Cognition and Instruction* 13, 269-313.

Nemirovsky R.C. (1994) On ways of symbolizing: The case of Laura and the velocity sign, *Journal of Mathematical Behavior* 13, 389-422.

Rabardel P. (1995) *Les hommes et les technologies - approche cognitive des instruments contemporains* [Men and technologies - a cognitive approach to contemporary instruments]. Paris: Armand Colin.

Rabardel P. & Samurçay R. (2001) From Artifact to Instrumented-Mediated Learning, *New challenges to research on learning*, International symposium organized by the Center for Activity Theory and Developmental Work Research, University of Helsinki, March, 21-23.

Roth W. & Tobin K. (1997) Cascades of inscriptions and the re-presentation of nature: how numbers, tables, graphs, and money come to re-present a rolling ball, *International Journal of Science Education* 19(9), 1075-1091.

Sfard A. (1991) On the dual nature of mathematical conceptions: Reflections on processes and objects as different sides of the same coin, *Educational Studies in Mathematics* 22, 1-36.

Tall D. & Thomas M. (1991) Encouraging versatile thinking in algebra using the computer, *Educational Studies in Mathematics* 22, 125-147.

Thompson A.G., Philipp R.A., Thompson P.W. & Boyd B. (1994) Calculational and conceptual orientations in teaching mathematics, *Yearbook of the National Council of Teachers of Mathematics*. Reston, VA: National Council of Teachers of Mathematics.

Trouche L. (2000) La parabole du gaucher et de la casserole à bec verseur: étude des processus d'apprentissage dans un environnement de calculatrices symboliques [The parable of the left-hander and the pouring pan: Study of learning processes in a symbolic calculator environment], *Educational Studies in Mathematics* 41, 239-264.

Vergnaud G. (1987) About constructivism, a reaction to Hermine Sinclair's and Jeremy Kilpatrick's papers, in J. Bergerson, N. Herscovics & C. Kieran (Eds.), *Proceedings of PME 11* (Vol. 1, pp. 73-80). Montreal: University of Montreal.

Vergnaud G. (1996) Au fond de l'apprentissage, la conceptualisation [At the base of learning, conceptualisation], in R. Noirfalise & M.J. Perrin (Eds.), *Actes de l'école d'été de didactique des mathématiques* (pp. 174-185). Clermont-Ferrand: IREM, Université de Clermont-Ferrand II.

Verillon P. & Rabardel P. (1995) Cognition and Artefacts: A contribution to the study of thought in relation to instrumented activity, *European Journal of Psychology of Education* 10, 77-103.

Vygotsky L.S. (1978) *Mind in society: The development of higher psychological processes*. Cambridge, MA: Harvard University Press.

Wenger E. (1998) *Communities of practice: Learning, meaning, and identity*. New York: Cambridge University Press.

Wenger R.H. (1987) Cognitive science and algebra learning, in A. Schoenfeld (Ed.), *Cognitive science and mathematical education*. Hillsdale, NJ: Lawrence Erlbaum Associates.

Yackel E. & Cobb P. (1996) Sociomathematical norms, argumentation, and autonomy in mathematics, *Journal for Research in Mathematics Education* 27, 458-477.

Chapter 8

INSTRUMENTAL GENESIS, INDIVIDUAL AND SOCIAL ASPECTS

Luc Trouche
LIRDEF, LIRMM & IREM
Université Montpellier II, France
trouche@math.univ-montp2.fr

Abstract: In Chapter 6, we analyzed didactic phenomena occurring during experiments in integrating symbolic calculators. We then showed how adopting an instrumental approach to analyzing these phenomena helped in understanding the influence of such tools upon mathematical activity and upon knowledge building. It is during the process of instrumental genesis that a calculator becomes a mathematical instrument.

In the first part of this chapter, we analyze the different forms that instrumental genesis takes by studying students' behavior so as to establish a typology of work methods in calculator environments. This typology indicates that the more complex the environment, the more diverse the work methods, and, consequently, the more necessary the intervention of the teacher in order to assist instrumental genesis.

In the second part of this chapter, taking this necessity into account, we introduce the notion of instrumental orchestration, defined by a didactical configuration and its modes of exploitation.

An orchestration is part of a scenario for didactical exploitation which aims to build, for every student and for the class as a whole, coherent systems of instruments.

Key words: Command process, Instrumental orchestration, Metaknowledge, Scenario in use, Work methods.

1. DIFFERENCES IN INDIVIDUAL INSTRUMENTAL GENESIS

Students have very different relationships with their calculator. Several methods help to pinpoint this diversity: conducting surveys of a wide student population to elicit their answers to a few questions posed at a given time, or following the instrumental genesis for a few students over the course of quite a long period of time. These different methods make it possible to bring out, for students' behaviors, several 'types of typologies'.

1.1 Local typologies

These typologies take into account only some aspects of instrumental genesis. In this sense, we speak of local typologies.

1.1.1 A typology linked to calculator learning type

Faure & Goarin (2001), from a survey of 500 10th grade students (most of them using graphic calculators), propose a typology depending on *the calculator learning type*. They take into account three approaches: learning with the teacher, learning through instructions for use, and learning by trial and error. Then the authors distinguish, within the given population, four profiles (Figure 8-1):

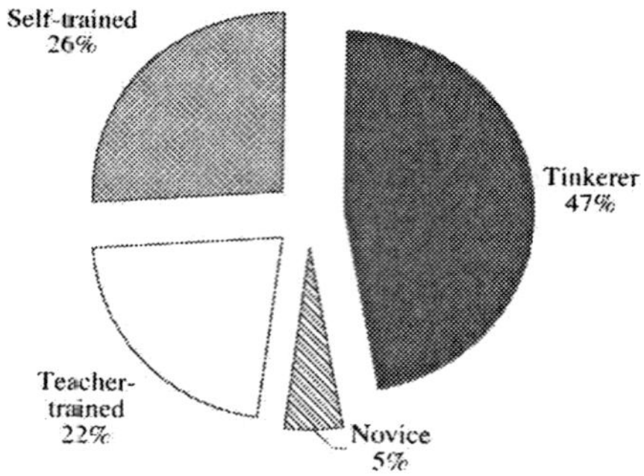

Figure 8-1. Distribution of calculator learning types

- *teacher-trained students* (22%) who have primarily learnt calculator use from the teacher;
- *self-trained students* (26%) who have primarily learnt calculator use from the instructions for use, and not substantially from the teacher;
- *tinkerers* (47%) who have learnt calculator use without guidance from any institutional source (whether teacher or instructions for use);

- *novices* (5%) who have had no training from the teacher, not consulted the instructions for use, and not tried to learn by themselves.

This typology can be related to some questions asked to the students.

i) "Do you know, with your calculator, how to find an approximate value for $\frac{\pi - \sqrt{2}}{\sqrt{2} + 1}$, to define a given function, to use a table of values, to graph a function, to choose an adequate window, to write programs?"

We can see, Figure 8-2, the frequencies for the answers *well* and *very well.*

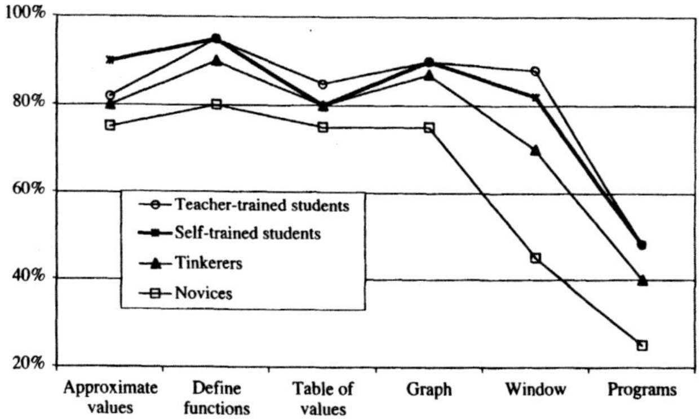

Figure 8-2. Typology and type of calculator knowledge

Techniques related to function graphs seem to be best mastered by all students. Teacher-trained students and self-trained students appear very confident in all the domains, whereas novices, logically, show quite weak competencies.

ii) "Is a calculator *useful* for computation, studying function variation, finding function limits, graphing functions, solving equations, and studying the domains for which functions are defined?"

Figure 8-3 shows the frequencies of the response *very much*:

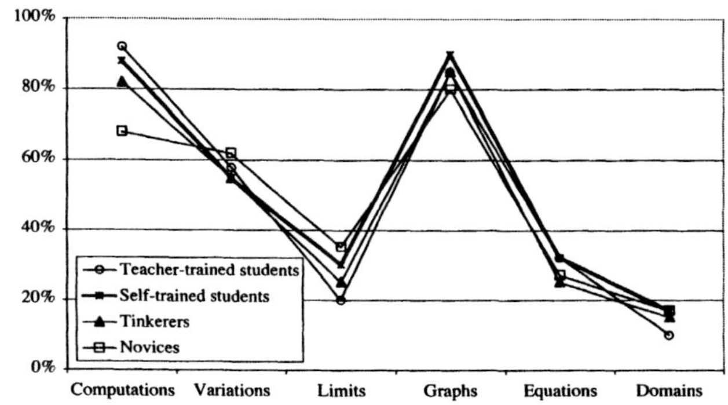

Figure 8-3. Typology and calculator usefulness

Answers for the different profiles are very similar: for all students, calculators are used essentially for computing and graphing functions.

iii) "Is a calculator useful *in the classroom* (for assessment, for the lesson, to help research), *at home* (for exercises, to learn lessons, to explore)?"

Figure 8-4 shows the frequencies of the response *very much*:

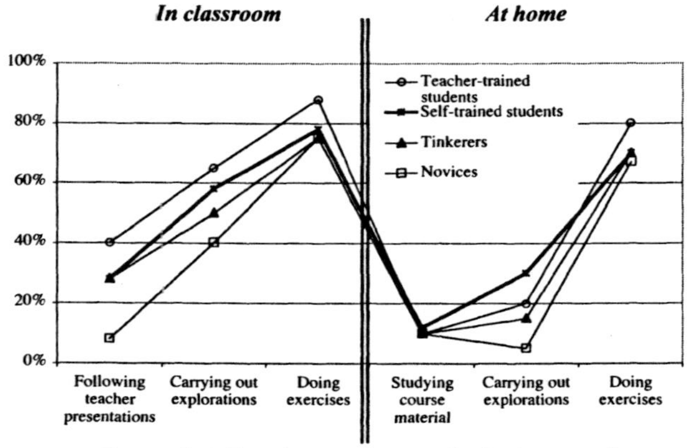

Figure 8-4. Typology and type of calculator work

Teacher-trained students accord a greater importance to calculator use during the lesson, which is to be expected: if the teacher has shown them how to use a calculator, s/he probably uses it in her/his mathematics teaching.

iv) "What is the relative importance of your *notebook*, your *textbook* and your *calculator*?"

Figure 8-5 shows the frequencies of the responses *great* and *essential*:

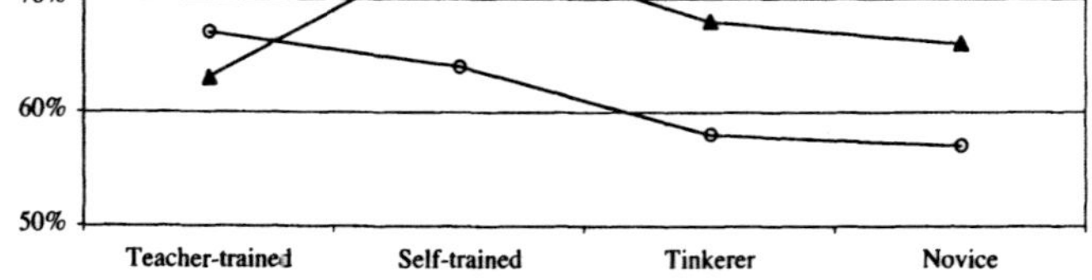

Figure 8-5. Typology and usefulness of mathematical tools

In each case, the notebook appears the essential tool, but, for trained students, calculator use overtakes textbook use. It is also interesting to remark that self-trained students, who learned calculator use from the printed instructions for use, are the most frequent users of textbooks.

To summarize, for the four questions asked, while the typology gives some results it does not identify great differences in calculator use. Some elements can explain this situation:

- the nature of the different typology categories is not the same: the teacher-trained students have not chosen to be trained (they have simply been in a classroom where a teacher took charge of this training), whereas self-trained students, tinkerers and novices are related by being placed in a situation (of no institutional training) which imposes a personal choice (choosing to learn from written instructions for example rather than through a trial and error strategy);
- the nature of tool utilization does not depend only on the type of training. Instrumental genesis is a process: other elements necessarily intervene (mainly individual work methods and learning environments, which are obviously different for the 500 students surveyed).

1.1.2 A typology linked to the privileged frame of work

Defouad (2000) notes that "instrumental genesis is not the same for all students; it depends on their personal relationships with both mathematics and computer technologies". He adds that, at the beginning of instrumental genesis, the relationships with graphic calculators are the most important. He distinguishes thus, in this graphic calculator environment, three profiles, a *numerical* one, a *graphical* one and a *paper-and-pencil* one, according to the frame of work privileged by the student: computation by calculator for the numerical profile, graphing with calculator for the graphical one, and obviously, work mainly with paper and pencil for the paper-and-pencil profile.

These categories are not stable:

- Defouad shows that, over the course of instrumental genesis, the nature of students' relationships with mathematics becomes more and more influential;
- we saw (Figure 6-10) that the applications employed to achieve tasks could change, according to the type of environment (for example, when moving from a graphic calculator environment to a symbolic calculator environment).

1.2 A more global typology

1.2.1. Principles for a typology

Understanding differences in students' behaviors is quite difficult. It needs to take into account, over a long time, more than just their privileged frame of work. For example, Mouradi & Zaki (2001) took into account the importance of paper-and-pencil work, but also the knowledge that students effectively use, interactions between pairs of students, students and teacher, and finally between students and computer. We have also (Trouche 2000) tried to consider various elements:

- *information sources* used, which can be previously built references, resort to paper-and-pencil, to the calculator or to the setting (in particular, during research activity, in practical work for example);
- *time of tool utilization* (both the global time over which the calculator is in use, and the time spent performing each instrumented gesture);
- *relationship of students to mathematics* and in particular *proof methods*: proof can proceed through *analogy*, *demonstration*, *accumulation of corroborating clues* (a particular form of over-checking, Chapter 6, § 1.2.2), from *confrontation* (based on comparison of various results obtained via the different information sources), and last from *cut and paste* (based on the transposition of isolated and not necessarily relevant pieces of proof);
- *metaknowledge* that is to say, knowledge which students have built about their own knowledge (Box 8-1).

Box 8-1.

Metaknowledge

Somebody is never in a wholly 'new' situation when discovering an artifact. S/he has already built knowledge about her/his environment and about her/himself, which is to say *metaknowledge*. Metaknowledge has emerged from several research fields:

- in the field of Artificial Intelligence, Pitrat (1990, p.207) distinguishes, between metaknowledge, knowledge about knowledge, knowledge about one's own knowledge, knowledge necessary to manipulate knowledge;
- in the field of didactics of mathematics, Robert & Robinet (1996) distinguish knowledge linked to mathematics, knowledge linked to gaining access to mathematical knowledge, and knowledge about one's own mathematical functioning (here these authors evoke the notion of control, as a global metaknowledge);
- in the field of cognitive psychology, Houdé & al (2002) also raise the question of control, when speaking of the co-existence, in each person, of both relevant and non-relevant schemes. If rationality, which generally exists for each individual, doesn't appear in her/his cognitive performances, the reason is often that the irrelevant schemes have not been inhibited.

We have stressed (Trouche 2000) the central role of the subject's *control* of her/his own activity. More precisely, we named this control *command process,* defined as the "conscious attitude to consider, with sufficient objectivity, all the information immediately available not only from the calculator, but also from other sources, and to seek mathematical consistency between them" (Guin & Trouche 1999b).

This command process takes place within a chart of essential knowledge (Figure 8-6), which is required in mathematical activity, in particular, when using symbolic calculators. It distinguishes two types of metaknowledge:

- first-level metaknowledge which makes it possible to seek information (*investigation*) from several sources: built references -- both material and psychological --, paper-and-pencil, the calculator, other students -- in particular within group work -- which makes it possible to store this information or to express it;
- second-level metaknowledge which makes it possible to process this information (*semantic interpretation, inference, coordination-comparison* of information coming from one or several sources, from one or several calculator applications or from other students).

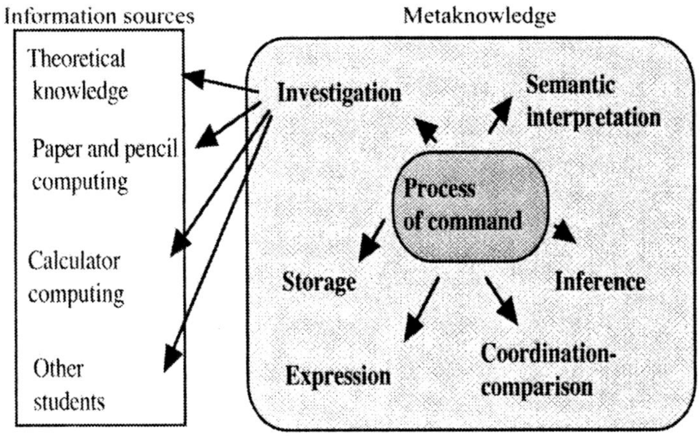

Figure 8-6. Information sources and metaknowledge

This chart itself does not completely describe a subject's behavior:

- each aspect of metaknowledge should be more clearly defined; for example investigation does not have the same character if it is applied only to the calculator or to the textbook, or to the setting as against it unfolding in all directions; the storage of new knowledge can be achieved alongside former knowledge or can lead to a *cognitive reorganization* (we know from Dorfler (1993) that experiences in computerized environments do not easily lead to such a cognitive reorganization);

- it is necessary to give a more precise description of the order in which the different types of metaknowledge are made use of, the respective time attributed to each of them. This precise description could be given when one has to describe the action of a given subject aiming at executing a given task in a given environment. The chart above could then provide us with a grid for analyzing this action.

Considering these different points, we have identified five extreme types of behavior, from observing students' work over a whole year (firstly in a graphic calculator environment, then in a symbolic calculator environment) and from analysis of their written productions and questionnaires, regularly handed out to them:

- a *theoretical* work method[1], characterized by the use of mathematical references as a systematic resource. Reasoning is based essentially on analogy and over-excessive interpretation of facts with occasional use of calculator;
- a *rational* work method, characterized by reduced use of a calculator, and mainly employing a traditional (paper-and-pencil) environment. What is distinctive here is the strong command by the student, with inferences playing an important role in reasoning;
- an *automatistic* work method, characterized by similar student difficulties whether in the calculator environment or in the traditional paper-and-pencil environment. Tasks are carried out by means of cut and paste strategies from previously memorized solutions or hastily generalized observations. The rather weak command by the student is revealed by trial and error procedures with very limited reference to understanding of the tools used, and without strategies for verifying machine results;
- a *calculator-restricted* work method, characterized by information sources more or less restricted to calculator investigations and simple manipulations. Reasoning is based on the accumulation of consistent machine results. Command by the student remains rather weak, with an avoidance of mathematical references;
- a *resourceful* work method, characterized by an exploration of all available information sources (calculator, but also paper-and-pencil work and some theoretical references). Reasoning is based on the comparison and the confrontation of this information, involving an average degree of command by the student. This is revealed in the form of investigation of a wide range of imaginative solution strategies: sometimes observations prevail, at other times theoretical results predominate.

The time devoted to each instrumented gesture is also an important element when discriminating between the various types of work method:

- this time is extremely brief for calculator-restricted and rational work methods. In the first case, because *zapping* behavior (Chapter 6, § 1.2.2) involves going from one image to another without thinking; in the second case, because the calculator is only used for targeted choices (no hesitation before nor adjustment after doing it);

- this time is much longer for theoretical, automatistic and resourceful work methods, for different reasons: for the theoretical and resourceful work methods, this time is necessary in order to analyze and compare one result with others; for the automatistic work method time is necessary to carry out the gesture itself and to understand the calculator result.

We summarize this typology in Figure 8-7.

Various sources of information were used to build this typology. Among them, practicals play an important role. Below we illustrate the typology in relation to a particular task studied during the course of the research.

Work method	Theoretical	Rational	Automatistic	Calculator restricted to	Resourceful
Privileged information source	Theoretical references	Paper and pencil	No single source	Calculator	No single source
Privileged metaknowledge	Interpretation	Inference	Investigation	Investigation	Comparison
Privileged proof method	Analogy	Demonstration	Copy and paste	Accumulation	Confrontation
Command process	Medium	Strong	Weak	Weak	Medium
Global time for calculator work	Medium	Short	Medium	Long	Medium
Time devoted to each instrumented gesture	Long	Short	Long	Short	Long

Figure 8-7. Five work methods in a calculator environment

1.2.2 Illustration of the typology

We proposed work on this task to an experimental 12th grade class, in a graphic calculator environment: students did not have at their disposal a *Limit* command as in a symbolic calculator environment (Appendix 6-2). This work took place after a detailed lesson on function limits, in particular about polynomials (Trouche 2000).

The function is defined by:

$P(x) = 0.03x^4 - 300.5003x^3 + 5004.002x^2 - 10009.99x - 100100.$

The questions are:

- "determine its limit at $+\infty$;
- determine a calculator window which confirms your result".

This type of function is well known to students. Here the difficulty comes from the distance between the four real roots (– 10/3, 10, 10.01 and 10000), which makes the choice of a relevant window difficult. On the standard window (Figure 8-8), the graph obtained is not easy to interpret. On a fitted window, the graph does not correspond to the students' idea of a limit at +∞ for a function.

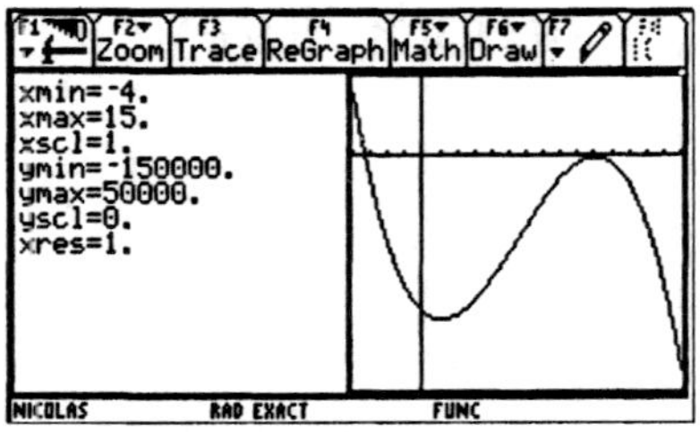

Figure 8-8. A function graphic, which looks quite strange

From observation of students it was possible to identify examples of the different work methods:

- *theoretical work method: student A*

S/he identified a polynomial function and evoked the relevant theorem: this polynomial has the same limit at +∞ as its term of highest degree. Therefore, $\lim_{+\infty} P$ is +∞. Through making a sketch on paper, A indicated that s/he knew the overall shape of a fourth degree polynomial.

This theoretical result was also used to find a relevant window for the calculator: A chose a wider and wider range for the x–axis [0, Xmax], adjusting Ymax with respect to the value of $0.03\,(\text{Xmax})^4$;

- *rational work method: student B*

B reproduced the method shown during the lesson (in order to demonstrate the theorem). So B factorized the term of highest degree. Then s/he gave the limits of each factor and, by applying the theorems on limit sums and products, found the limit of P. In order to obtain an appropriate window, B undertook a traditional function analysis: finding the derivative of P, then the derivative of P' and the sign of P''(x). Due to lack of time, B could not finish the work;

- *automatistic work method: student C*

C did not recognize a *reference* function form (Chapter 9, § 3.2.1, note 13) and used her/his calculator to form some idea about it. Defining the function on the calculator took rather a long time.

C was unable to analyze the graph shown on the standard window or to obtain a more appropriate window, and was not able to use a more theoretical approach, which seemed too difficult for such a complex object. The only information was obtained from the table of values (see right side). An answer was given from these results: $\lim_{+\infty} P$ seems to be $-\infty$;

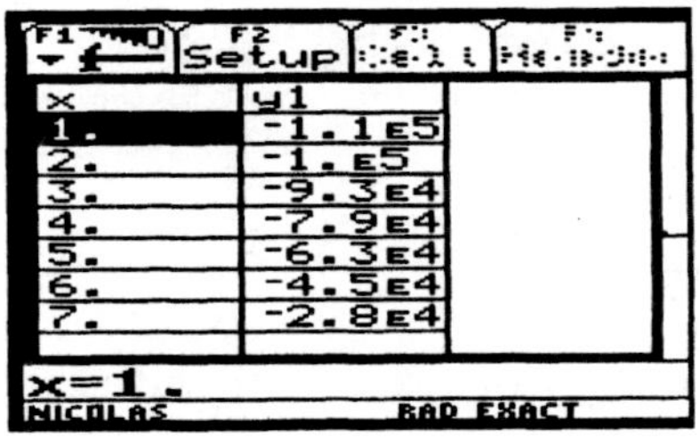

- *calculator-restricted work method: student D*

From the beginning, student D started looking for an appropriate window for the graph and carried out various tests involving numerous commands:

- the *Trace* command led to the location of points situated outside the screen and therefore, the student redefined the window to allow the overall graphic representation to be shown;
- secondly, *Zoom* commands permitted quicker searches.

D used all forms of exploration possible on this calculator (using the widest possible range for the x variable). In this way, s/he obtained the required result using only the resources of the calculator, without any reference to theoretical results, and without putting any record of her/his work on paper;

- *resourceful work method: student E*

Using theorems learnt during the lesson, E was able to assert that $\lim_{+\infty} P = +\infty$. Then s/he looked for confirmation through a graphic representation of the function. After some concordant tests, s/he assumed that the graphics invalidated her/his first result obtained with a theoretical argument: the function seemed to be strongly decreasing, even for high values of x. E therefore tried to solve this contradiction and to find a justification for the exceptional status of this function. Observing the expression of P, s/he noticed that the coefficient of x^3 was extremely large while the coefficient of x^4 was very small. For E this point justified the exceptional status of this polynomial:

- *"For standard coefficients, it is the term with the highest degree which counts;*
- *in this case (a great difference between these coefficients), it is the term x^3, which counts. Therefore $\lim_{+\infty} P$ is $-\infty$".*

This makes clear the characterization of this work method in terms of a search for coherence when confronted with various results from different sources.

This typology has also been put to test in other situations, particularly in symbolic calculator environments (Guin & Trouche 2002).

1.2.3 Interest of the typology

Clearly this typology does not aim to (and could not) constitute a partition of the work methods of different students in a given environment. The work methods of most students cannot be classified as one of these types: they generally fall between positions and they can move between one and another. However, this typology does make it possible to establish a geography of the class, which has three-fold interest:

- it gives indicators to mark out, at a given moment, the relationship of a student with the five working styles brought to the fore. Besides, these five poles appear in similar form in other work: Hershkowitz & Kieran (2001) distinguish, for example, two types of behavior, linked to different methods of coordinating representations in a graphic calculator environment: "A *mechanistic-algorithmic way* (where students combine representatives in non-thinking, rote ways) and a *meaningful way*". The former one is close to our calculator-restricted work method. The latter one looks like the theoretical work method, which we have previously described;
- it helps the teacher to play on the complementarities of the various work methods: we have shown (Trouche 1996) the interest of the association between rational and resourceful work methods for practicals. It thus gives the teacher a means to organize the learning environment. However, these evolutions depend significantly on work situations and arrangements set up by the teacher;
- it gives indicators to mark out the evolution of student approaches and thus to interpret their moves in terms of instrumental genesis. We have shown, for example (Trouche 1996) significant evolutions of the calculator-restricted work method toward the resourceful work method; we have also shown *that the more complex the environment, the more difficult the command process and the greater the diversity of work methods* (Trouche 2004), and, consequently, the more necessary the intervention of the teacher in order to *assist instrumental genesis* and *help the process by which the student exercises command.*

2. INSTRUMENTAL ORCHESTRATIONS

In both Chapter 6 and Chapter 8 (§ 1), the complexity of instrumental genesis is apparent, and so the need to articulate a set of instruments from a set of artifacts. Variability amongst students is also apparent in the instrumental genesis taking place in a given *class*[2]. Until now, we have considered these processes only in their individual dimension. But each utilization scheme has also a social dimension[3], of which Rabardel & Samurçay (2001) point out the importance:

> The world that genetic epistemology is interested in is a world of nature, not of culture. We have moved beyond this limitation by giving utilization schemes the characteristics of social schemes: they are elaborated and shared in communities of practice and may give rise to an appropriation by subjects, or even result from explicit training processes.

The integration of instruments within a class needs to take the process of instrumental genesis into account. Obviously while it does not remove the individual dimension of this process, it reinforces the social dimension, thus limiting individual diversity.

Box 8-2.

Didactic Exploitation System
(Chevallard 1992, p.195)

The successful integration of a technical tool in the teaching process requires complex and subtle work of *didactical implementation*. Chevallard uses a computer metaphor in order to distinguish three levels whose interaction is essential:
1) *Didactical hardware*: didactical environment components, various artifacts (calculators, overhead projectors, teaching software...), but also instructions for use, technical sheets, etc.
2) *Didactical software*: mathematics lessons.
3) *Didactic exploitation system*: essential level concerned with making relevant use of the potential resources of a didactical environment and with achieving both the coordination and integration of first and second levels.
Chevallard underlines the importance of this third level: without it, the didactical hardware components run the risk of being completely excluded from the teaching scene. Within the history of introducing computers into the institution of schooling, account has begun to be taken of the necessity of acknowledging and coordinating these three levels only in the face of failures and under the pressure of disappointments. Available software (word processing, spreadsheets, CAS, etc.) has not generally been conceived with teaching in mind and thus requires exceptional work for didactical implementation, along lines which have hardly been developed in rough.
A didactic exploitation system requires *didactical exploitation scenarios*. A didactical exploitation scenario is composed of a pedagogical resource and its implementation modes (in an ordinary classroom, in a special classroom, etc.) referred to as a *didactical configuration*.
Solving problems of the didactical integration of computerized tools requires the development of a true *didactical engineering* of computerized tools. The didactical engineer, between computer scientist and teacher, does not yet exist (or only as a prototype). Once such

a profession does exist, the teacher will be freed from tasks which s/he cannot carry out (didactical materials production) and will be able to become a specialist in teaching. *Further from machines; closer to students.* More than a teaching evolution, this would be a teaching revolution.

The *institution of schooling* has to take charge of these 'explicit training' processes. These explicit training processes require that a didactic exploitation system be designed (Box 8-2) so as to ensure the integration of tools in a class, and their *viability*.

This design requires models and exemplars of use:

> The degree to which this [CAS] technology is likely to be productive in the classroom will be highly dependent on the availability of proven models and exemplars to guide teachers and students in its use (Ruthven 1997).

Models and exemplars of use must include questions of management of time and space, and organization of tools within the classroom.

In order to take account of this necessity, we have introduced the notion of *instrumental orchestration* (Box 8-3) to refer to an organization of the artifactual environment, that an institution (here the schooling institution) designs and puts in place, with the main objective of *assisting* the *instrumental genesis of* individuals (here students).

An orchestration is part of a didactical exploitation scenario: it is designed in relation both a given environment and to a mathematical *situation* (Brousseau 1997). As states Rabardel (2001): "activity mediated by instruments is always *situated* and situations have a determining influence on activity".

Box 8-3.

The word *orchestration* is often used in the cognition literature. Dehaene (1997) uses this word pointing out an *internal* function of coordination of distributed neural networks. Ruthven (2002) also uses this word, in the mathematical field, pointing out a cognitive internal function (in relation to the construction of the derivative concept): "unifying ideas are careful orchestrations of successive layers of more fundamental ideas around a more abstracted term". In fact, the necessity of orchestrations, in this sense, clearly manifests itself in the learning of mathematical sciences seen as "the construction of a *web* of connections - between classes of problems, mathematical objects and relationships, real entities and personal situation-specific experiences" (Noss & Hoyles 1996, p.105). In our sense, the word orchestration means an *external* steering of student's instrumental geneses.

The word *orchestration* is indeed quite natural when speaking of a set of *instruments*.

The *orchestration*, in the musical register, may indicate two things:

- the work of the composer to adapt, for an orchestra, a musical work originally written for only one instrument or a few;

- "*the art to put in action various sonorities of the collective instrument* which one names orchestra by means of infinitely varying combinations" (Lavignac, French musicographer, 1846-1916).
By choosing this word, we refer here to this second and more general sense.

Both define an *instrumental orchestration*:

- a set of *configurations* (i.e. specific arrangements of the artifactual environment, one for each stage of the mathematical situation);
- a set of *exploitation modes* for each configuration.

These exploitation modes may favor production of activity accounts. These accounts can themselves be integrated as new learning and teaching tools.

An instrumental orchestration may act *mainly* at several levels:

- at the level of the artifact itself;
- at the level of an instrument or a set of instruments;
- at the level of the relationship a subject maintains with an instrument.

These levels correspond to the three levels of artifacts distinguished by Wartofsky (1983):

- "The level of primary artifacts which corresponds to the concept of the artifact as it is commonly used (...), computers, robots, interfaces and simulators;
- (...) [The level of] secondary artifacts, which consists of representations both of the primary artifacts and of modes of action using primary artifacts;
- (...) The level of tertiary artifacts (...) represented, for trained subjects in particular, by simulated situations as well as by reflexive methods of self-analysis of their own or the collective activity".

2.1 A first level instrumental orchestration: guide to mathematical limit

We have proposed such a guide (Trouche 2001) to assist learning of the idea of limit.

In order to define this orchestration, we have to analyze the gap between the mathematical idea to be taught (Box 8-4) and the ways in which the artifact has implemented it (Box 8-5).

From these constraints of the artifact, one can generate some hypotheses about the techniques which students put in place to study limits of functions, and about the operational invariants (Boxes 6-4 & 6-5) likely to be built in such an environment. The constraints of the TI-92 do not favor moving beyond a kinematic point of view on the idea of limits:

Box 8-4.

The limit concept

Several *frames*[4] (geometrical, algebraic and numerical) are involved in studying limits, sometimes creating productive intuitions, sometimes acting as obstacles. Trouche (2003) distinguishes two main points of view:
- a *kinematic* point of view, within a generally geometrical frame: a quantity y (depending on x) tends towards b as x tends towards *a* if, when x becomes closer to *a*, y becomes closer to b. For this definition, movement has a crucial role: one can say that 'variable *pulls* function'. The geometrical frame is also important: the limit idea involves bringing together graphical representatives or geometrical objects (curves and asymptotes for example);
- an *approximation* point of view, in a numerical frame: a quantity y (depending on x) tends towards b as x tends towards a if y can be as close to b as one wants, as long as x is close enough to *a*. It is thus the degree of precision that one wants which *constrains* the variable.

Construction of the limit concept involves going beyond the first point of view, but it is often an *articulation* of the two points of view that permits this notion to be grasped.

- the calculator, through its symbolic application, only gives (if it 'knows' an answer) the value of the limit, which is not sufficient to give sense to the idea;
- through its graphical and numerical applications, the calculator clearly presents a kinematic point of view.

We want to define an orchestration aiming to support instrumental genesis, transforming the TI-92 artifact into an instrument for computation of limits. To achieve this objective, it is necessary to *fit out* the artifact itself (it is a first level orchestration) in order to favor, in this environment, the passage from a kinematic point of view to an approximation point of view.

Box 8-5.

Constraints on limit computations of one symbolic calculator (TI-92) and corresponding potentialities

We use here the typology of constraints presented in Chapter 6, § 2.1.

1) The *internal constraints* (what, by nature, the artifact can do?)
A symbolic calculator contains a CAS application (Computer Algebra System); it can determine an exact limit provided that the corresponding "knowledge" has been entered.
A symbolic calculator can also (as a graphic calculator) give graphic or numerical information on the local behavior of functions. The processing is done by numerical computation.

2) The *command constraints* (what are the available commands?)
Only one *Limit* keystroke exists. It is a formal command, located in the calculator symbolic application.

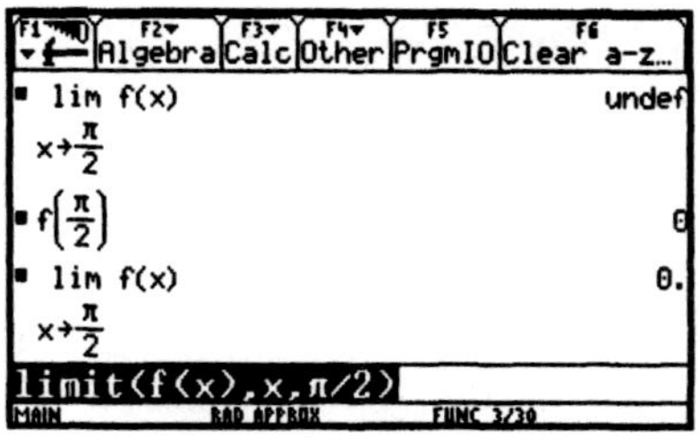

Its syntax is (see screen) "limit(f (x),x,a)"; it corresponds to the order of the statement "the limit of f(x) as x tends toward a". Nevertheless this command can be combined with the *approximate detour* (Chapter 6, § 2.3.1). In the example shown, the *Limit* command, applied to the function $f(x) = (\cos x)^x$ does not give the result "directly", but gives it by switching to approximate detour (screen copy, 3rd line).

3) The *organization* constraints (how are the available commands organized?)

The different applications (symbolic, graphical or numerical) permitting the study of functions are directly accessible on the keyboard. As a part of graphical or numerical applications, the operation of the calculator requires the interval of x and then the interval of y to be chosen first. This is a natural order for the study of functions, but is not an adequate order for studying limits (Box 8-4): the mathematical organization and tool organization are opposed from a chronological point of view.

This configuration rests on putting in place, within the calculator, three levels for each study of a limit. We present these through an example of a limit, the value of which is not known by the calculator $\lim_{x \to +\infty} \frac{\sqrt{x} + \cos x}{x + \sin x}$ (Appendix 6-2).

These three levels[5] are accessible from a subsidiary menu linked to the *Limit* command (Figure 8-9).

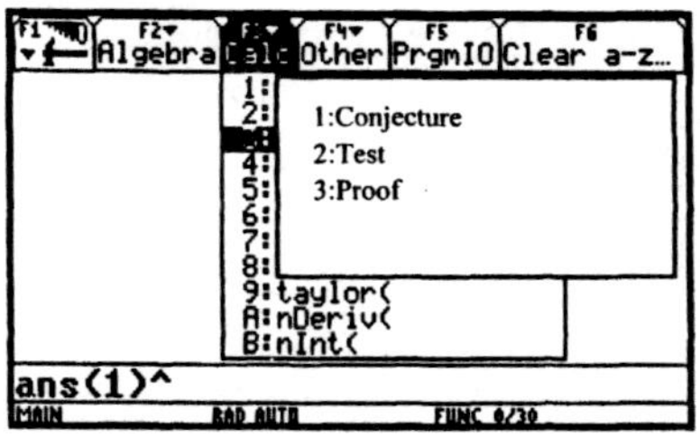

Figure 8-9. Menu and sub-menu allowing access to three levels of study of limits

Level 1. Conjecture searching

The *Conjecture* command gives access to a split screen: on the left side is the *TABLE of values* application, on right side, the *GRAPH* application (Figure 8-10). The split screen allows these two applications to be connected.

One has to choose a table setting and a graph window (here, the study being in the neighborhood of infinity, corresponding to 'large' values of the variable x).

Observation of both tables of values allows a conjecture to be formed: maybe the function limit is equal to 0.

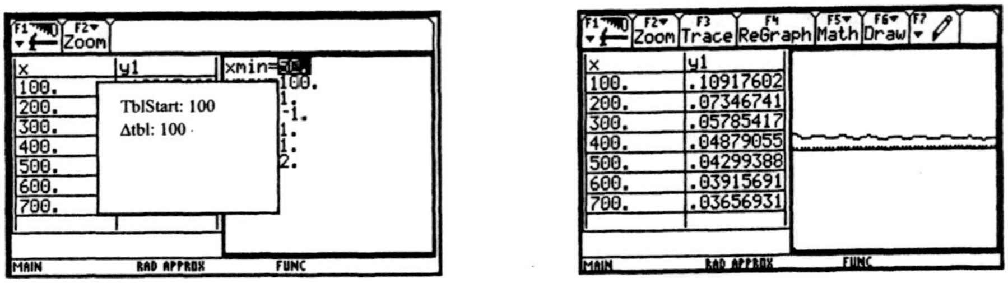

Figure 8-10. Conjecture searching

Level 2. Testing

The *Testing* command gives access to a new split screen, again with the table of values and graphical applications (Figure 8-11). But here there is a fundamental logical reversal: one has to choose first the neighborhood for y, image of the variable x.

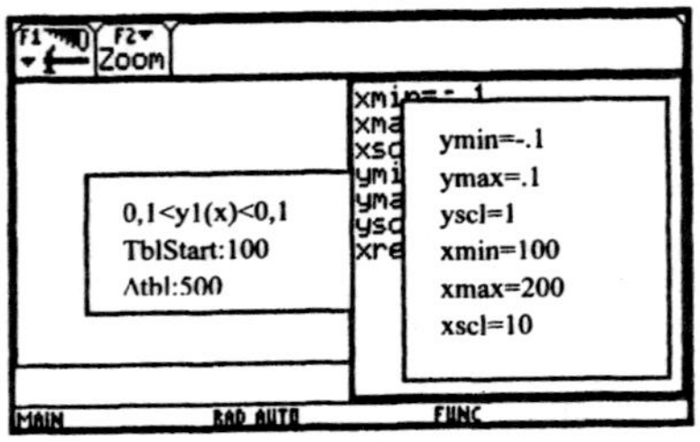

Figure 8-11. Conjecture testing

It is thus the degree of precision one wants which *constrains* the variable. It is a sort of challenge, linked to the approximation point of view (Box 8-4): if one wants y to be in a given neighborhood of 0, in which interval $[a, +\infty[$ is it sufficient to choose x?

The obtained numerical and graphical results (Figure 8-12) show that the constraints on the variable are not sufficient: the aimed degree of approximation is not achieved. The student thus has to go back to choose a new table set and graph window. These gestures are not only gestures of exploration: they are preparing the passage from a kinematic point of view to an approximation point of view.

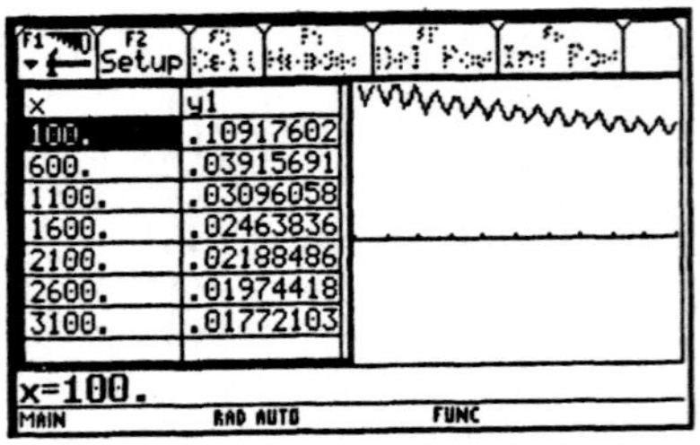

Figure 8-12. Not sufficient constraints on the variable

Level 3. Proof

The *Proof* command gives access to a new split screen: on the left side, the symbolic application, on the right side a work sheet dedicated to proof (Figure 8-13). The symbolic application can give the limit value (although this is undefined in the case shown here).

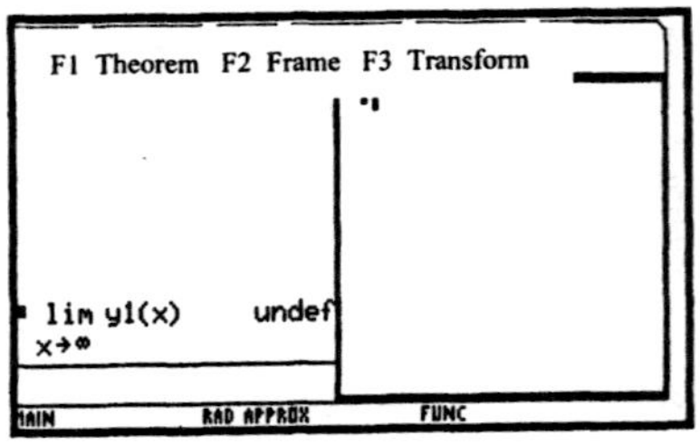

Figure 8-13. Proof screen

The right work sheet gives access to menus which allow the "guiding of thinking about problems and tool classification principles" (Delozanne 1994):

- the *F1* menu gives access to a set of theorems. It leads to work on characterizing the functions studied (for example, "is it possible to apply theorems on limits of rational functions to the given function?");
- the *F2* menu allows framing strategies to be tried out, with the help of calculator numerical and graphical applications. It leads to a comparison with *reference functions* (Chapter 9, § 3.2.1, note 13);
- the *F3* menu gives access to symbolic functionalities (factoring, expanding, etc.).

We have thus defined an orchestration *configuration*. Defining orchestration involves choosing some *exploitation modes*. Several modes are possible:

- this limit math guide can be always available or available only during a specific teaching phase;
- students can be free to use this guide, when available, as they want, or they can be obliged to follow the order of the three given levels;
- the list of stored theorems can be fixed, or it can be progressively established, linked to the mathematical lessons, built in the classroom and collectively stored in each calculator;
- activity accounts can be required, describing all the steps of instrumented work, or not.

The orchestration defined in this way, modifying the artifact itself, is not a matter of building the explication module of an expert system (besides, Delozanne (1994) indicates that this task is quasi-impossible if the software designer has not initially taken this development into account). It could only constitute a *guide*[6], an assistant for instrumental genesis, in the study of limits, making it possible to move from one frame to another, and providing balance between the two points of view constituting this notion (Box 8-4). Designing such an orchestration involves analyzing precisely both the notion to be taught (from an epistemic point of view) and the way in which the artifact has implemented it. It does not solve, in itself, the problem of the learning process of the limit idea: one also has to choose a field of critical functions, more generally a field of problems nested in didactic situations which have to be elaborated. Defining a didactical exploitation scenario requires then the choosing of an orchestration which is well adapted to each stage of this problem treatment.

2.2 A second level orchestration: around the sherpa-student

The utilization of individual tools within the school, in the form of calculators fitted with a small screen, raises the issue of the socialization of students' actions and productions. This socialization requires particular arrangements. Since the beginning of the 1990s, there has been a particular artifact -- a view-screen -- which allows one to project the calculator's small screen[7] onto a big screen, which the entire class can see. Guin & Trouche (1999a) presented an instrumental orchestration, which exploits this arrangement with the main objective of socializing -- to a certain extent -- students' instrumental genesis.

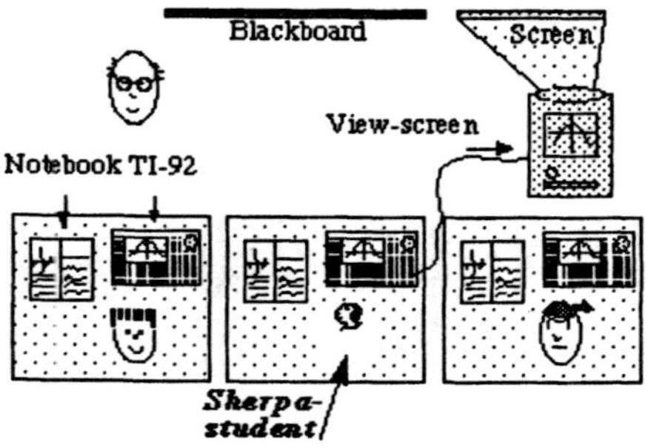

Figure 8-14. The sherpa-student, part of an instrumental orchestration

The *configuration* of this orchestration (Figure 8-14) rests on the devolution of a particular role to one student: this student, called the *sherpa-student*[8], pilots the overhead-projected calculator. S/he will thus be considered, for both class and teacher, as a reference, a guide, an auxiliary and a mediator. This orchestration favors the collective management of a part of the instrumentation and instrumentalization processes (Chapter 6, § 2.2): what a student does with her/his calculator -- traces of her/his activity -- is seen by all. This allows one to compare different instrumented techniques and gives the teacher information on the schemes of instrumented action being built by the sherpa-student.

It also has other advantages:

- the teacher is responsible for guiding, through the student's calculator, the calculators of the whole class (the teacher does not perform the instrumented gesture but checks how it is performed by the sherpa-student). The teacher thus fulfils the functions of an *orchestra conductor* rather than a *one-man band*[9];

- for his/her teaching, the teacher can combine paper-and-pencil results obtained on the board, and results obtained by the sherpa-student's calculator on the class screen. This facilitates, for students themselves, the combination of paper-and-pencil work and their calculator work at their own desk.

Several *exploitation modes* of this structure may be considered. The teacher may first organize work phases of different kinds:

- sometimes calculators are shut off (and so is the overhead projector): it is then a matter of work in a paper-and-pencil environment;
- sometimes calculators are on as well as the overhead projector and work is strictly guided by the sherpa-student under the supervision of the teacher (students are supposed to have exactly the same thing on their calculator-screens as is on the big screen for the class). Instrumentation and instrumentalization processes are then strongly constrained;
- sometimes calculators are on as well as the overhead projector and work is free over a given time. Instrumentation and instrumentalization processes are then relatively constrained (by the type of activities and by referring to the sherpa-student's calculator which remains visible on the big screen);
- sometimes calculators are on and the projector is off. Instrumentation and instrumentalization processes are then only weakly constrained.

These various modes seems to illustrate what Healy (2002) named *filling out* and *filling in*[10], in the course of classroom social interaction:

- when the sherpa-student's initiative is free, it is possible for mathematically significant issues *to arise out* of the student's own constructive efforts (this is a filling out approach);
- when the teacher guides the sherpa-student, it is possible for mathematically significant issues to be *appropriated during* children's own constructive efforts (filling in approach).

Other variables must also be defined: will the same student play the role of the sherpa-student during the whole lesson or, depending on the results proposed, should such and such a student's calculator be connected to the projector? Must the sherpa-student sit in the front row or must she/he stay at her/his usual place? Do all students have to play this role in turn or must only some of them be privileged?

Depending on the didactic choices made, secondary objectives of this orchestration can arise:

- to favor debates within the class and the making explicit of procedures: the existence of another point of reference distinct from the teacher allows new relationships to develop between the students in the class and the teacher, between this student and the teacher -- about a result, a conjecture, a gesture or a technique --;
- to give the teacher means through which to reintegrate remedial or weak students into the class. The sherpa-student function actually gives remedial

students a different status and forces the teacher to tune his/her teaching procedures to the work of the student who is supposed to follow her/his guidelines. Follow-up of the work by this student shown on the big work-screen allows very fast feed-back from both teacher and class.

This instrumental orchestration involves coordination of the instruments of all students in the class and favors the connection, by each individual, of different instruments within her/his mathematics work.

2.3 Another second level orchestration: practicals

Guin & Trouche (1999a) present an organization of students' research work in a calculator environment: *practicals.*

This orchestration aims to:

- make it possible for instrumental genesis to proceed at its own rhythm;
- develop social interactions between peers;
- favor establishment of relationships between different tools (calculator and paper-and-pencil) within a research process.

The configuration is this one: each student has at her/his disposal a calculator, paper-and-pencil. Students work in pairs (work groups are small, because of the smallness of the calculator screen) to solve a given problem. These problem situations (Appendix 8-1) are created with the aim of promoting interaction between calculators, theoretical results, and handwritten calculations as an aid to conjecture, test, solve and check. After working on these problem situations, each pair has to explain and justify their observations or comments, noting discoveries and dead-ends in a written research report. The role of this report is twofold:

- it focuses the student activity on the mathematics and not on the calculator, forcing students to give written explanations for each stage undertaken in their research (a very important step);
- it gives the teacher a better understanding of the various steps of the students' work method, and makes it possible to follow the instrumental genesis of students.

There is only one notebook for each pair. This choice is an important one: each research team is thus obliged to find a consensus, or to explain divergences.

Several *exploitation modes* are possible:

- students can be free (or not) to form themselves into pairs. We showed (Trouche 1996) the value of some specific pairings, for example a student with a quite *rational* work method and a student with a quite *calculator-restricted* work method: the interaction allows an evolution of each work method and some enrichment of instrumentation processes;

- students can be free (or not) to choose which one will write the research report;
- the teacher can offer appropriate assistance to help students out of deadlocks, to reinitiate reflection during the practical, or only at the end of it, or a week after;
- written research reports can be given to the teacher at the end of practicals, or a week later. In the first case, research reports are more faithful (showing what happens during practical, moment by moment, step by step). In the second case, students can have more time to read their own report, to think about their own work, to criticize their own research;
- after reading students' research reports, the teacher can give a problem solution, in relation to the students' results, or give only some partial indications opening up new strategies for students to pursue during further practicals.

In the frame of this orchestration, teachers and students play a new role, as stated by Monaghan (1997): "the teacher is viewed as a technical assistant, collaborator, facilitator and as a catalyst, and students have to cooperate in group problem solving".

2.4 A third level orchestration: mirror-observation

In the previous orchestration (§ 2.3), a research notebook constitutes an essential tool for students (for making explicit their own calculator and paper-and-pencil approaches, evaluation of the relevance of results, etc.). This notebook thus appears to be a tool for *activity self-analysis* (Rabardel & Samurçay 2001). We have presented (Trouche 2003) another arrangement, so called *mirror-observation* (Figure 8-15).

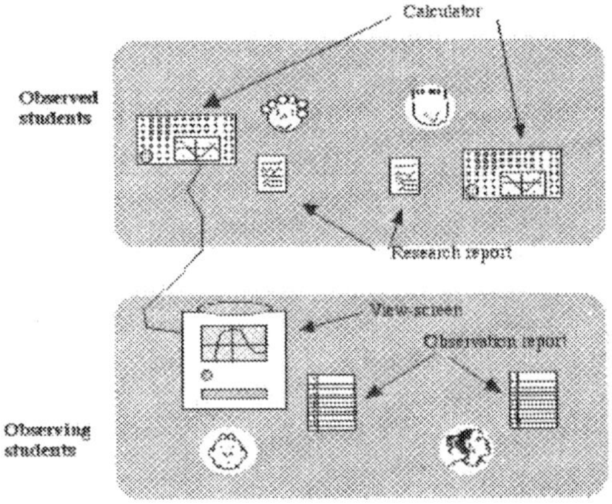

Figure 8-15. The *mirror-observation* configuration

This device aims at analysis of a work period during the course of the year. The *configuration* is this one: students work in pairs. Half of the pairs (Figure 8-15, *observed students*) carry out a given mathematical task. The other half observe and note the actions carried out, with the help of two artifacts:

- a palette for overhead projection makes it possible to capture the calculator screen of one of the students from the pair (the one who is not in charge of the written research report);
- observation sheets are used to note, every fifteen seconds[11], the whole of the students' actions. These sheets (Figure 8-16) appear as grids in which are located different types of tasks: *paper-and-pencil* tasks, *calculator* tasks (distinguishing the different applications involved), tasks relating to *interactions* (with the teacher, other students, or oneself: *hazy gaze*) and last the *other* actions that have nothing to do with the problem dealt with.

		15	*30*	*45*	*1mn*	*15*	*30*	*45*	*2mn*
Paper-and-pencil	Reading text of problem	x							
	Reading lesson								
	Reading accomplished work								
	Reading neighboring work								
	Drawing								
	Computation							x	x
Calculator	Machine condition (on/off)		x	x	x	x	x		
	Computation		x	x					
	Y Editor				x				
	Graph								
	Table of values								
	Neighboring machine								
Interactions	Hazy gaze					x	x		
	With teacher								
	With neighbor								
Other									

Figure 8-16. A timed observation sheet
(During the first 15 seconds, student reads problem text, then s/he uses her/his calculator for some computations, etc.)

Examination of the grids, corresponding to observation of five student pairs during the first five minutes is presented in chronological order (Figure 8-17). One may observe the large dispersion of work methods: pair number 1 takes a very short time reading the problem text, and rushes toward the calculator. Pairs number 3 and 4 spend a certain time on various irrelevant actions.

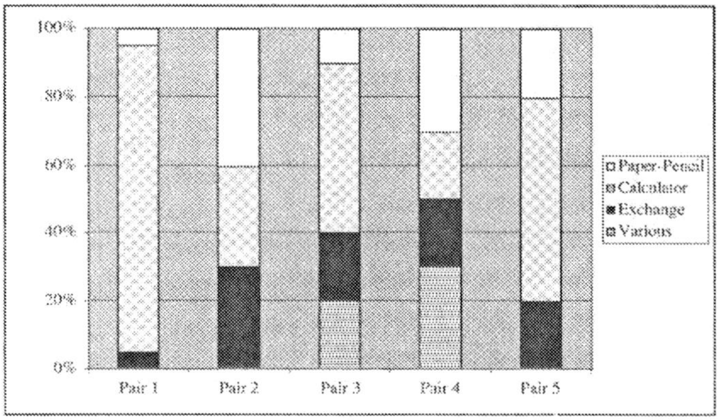

Figure 8-17. Compared synthesis of actions performed by the five pairs of students during the first five minutes of the activity

This instrumental orchestration highlights, not so much the results of the activities, but the various forms that these can take. It enables students, as they themselves noticed, *"while observing others to observe oneself"* (hence the suggestion of the term *mirror* for this type of observation) offering the possibility of an auto-analysis of the action, the construction by a *reflexive mediation*: for the majority of the observed subjects, the organization of action revealed by the chronological synthesis caused major surprise (for instance: *"How come, I haven't spent more than 15 seconds reading the text of the problem?"*). The gap between what the students actually did and what they remembered doing, as well as the gap between the written traces of the research report and the written traces of the observation sheet, allow a profound reflection on the shape of activity, allow the understanding of certain defaults and the rectification of certain failings.

Various exploitation modes are possible. Among them:

- this orchestration may be used only exceptionally, or may be a regular tool for regulation of students' instrumented activity;
- one may fix, or not, the role of each observed student (for example one can be in charge of the calculator, the other in charge of the research report);
- the type of tasks noted on the timed observation sheet can be modified, in relation to the type of mathematical problem set;
- each observation sheet analysis can be done within the group of four students (the students observed and the two observing), or all these observation sheets can be made public with the whole class.

Other devices can arouse students' thinking about their own activity[12]: for example some experiments (Trouche 1998) incorporated some form of "barometer" of the integration of instruments, i.e. questionnaires asking students about their instrumented activity. These accounts, giving information about what students think about their own activity, are complementary to observations from their peers.

3. DISCUSSION

Common elements can be recognized within these various instrumental orchestrations: a favoring of interaction between students, the publication and use of accounts of activity (sherpa-student's screen, research reports, or timed observation sheets), thus giving the teacher means to understand and guide the instrumental genesis of students. Instrumental orchestration combines all these elements to reinforce the *social dimension* of instrumented action schemes and to assist students in the process of command.

The need for a strong process of command is linked to the practice of mathematics; mathematics seen as "a web of interconnected concepts and representations which must be mastered to achieve proficiency in calculation and comprehension of structures" (Noss & Hoyles 1996). It is also linked to the tools available.

The necessity of taking the tools of the environment into account is not new. Proust (2000) notes, for example, recurrent mistakes in Babylonian numerical texts, in computation involving numbers composed of more than five figures. Her hypothesis is that these mistakes came from sticking together two computations realized with a tool linked to the five fingers of the hand. More exactly, it is a matter of the bad *articulation* of two types of artifacts: artifacts for material computation, and for writing, fingers being involved in both types of gesture.

What is true for 'old' computation environments is all the more true in a computerized environment (Basque & Doré 1998). Very sophisticated artifacts such as those available in a symbolic calculator environment give birth to a *set of instruments*. The articulation of this set demands from the subject a strong process of command, allowing her/him to build coherent *systems of instruments*. As Rabardel (2000) notes, this is a crucial point:

> This question seems to us particularly crucial in view of the current context of technological expansion. What artifacts should be proposed to learners and how should we guide them in their instrumental genesis and through the evolution and adjustment of their systems of instruments?

Instrumental orchestrations seem to give some elements of an answer to this question. They take into account artifacts in the learning environment, at three levels (tool level, instrument level, meta-level). They take place (Figure 8-18) within a didactical exploitation scenario (Box 8-2). According to the metaphor, we could say that designing an orchestration requires a musical frame. The following chapter will treat this point, i.e. the design of mathematical situations and problems.

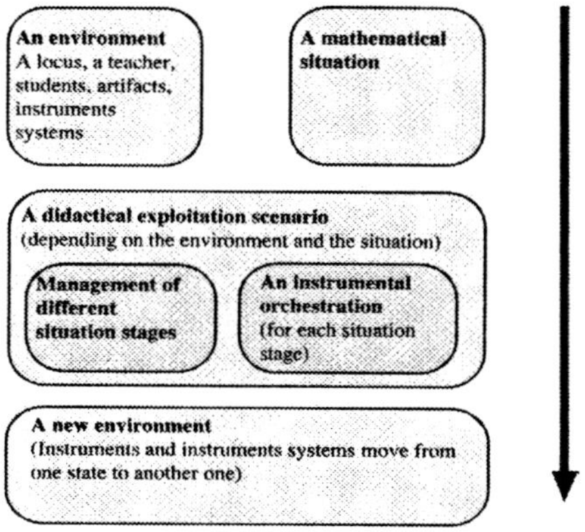

Figure 8-18. Evolution of a learning environment

After choosing a mathematical situation and defining the management of successive situation stages, designing an instrumental orchestration implies defining a didactical configuration and its exploitation modes.

A didactical exploitation scenario has effects on a learning environment:

- obviously it has effects on the knowledge built, via the treatment of mathematical situations;
- it has effects on the *didactical contract* (Box 8-6). For example, the devolution of a particular role to a sherpa-student (§ 2.2) enables parts of this contract to be made explicit;
- it has effects on instrumental genesis, i.e. on instruments and systems of instruments.

Box 8-6.

Didactical contract
(Brousseau 1997)

Brousseau evokes the *contract*, i.e. "the relationships determining - explicitly for a small part, but mainly implicitly, what each partner, teacher and learner, has to manage and what s/he will have the responsibility for". Part of this contract, which is related to the content, mathematical knowledge, is the *didactical contract*.
Brousseau points out the importance of the points at which this contract breaks down: "knowledge is precisely what will solve the crisis related to these breakdowns (...). The surprise of the student, when s/he can't solve the problem and rebels against the teacher who has not made her/him able to solve it, the surprise of the teacher who estimated her/his lessons to be sufficient... revolt, negotiation, search for a new contract depending on the new state of knowledge". The essential notion is therefore not the contract itself, but, through the breakdowns, the *process of searching for a hypothetical contract.*

Designing such scenarios of didactical exploitation is a complex task, calling for various competencies (Chevallard 1992): computer engineers, didactical engineers, curriculum designers, etc. Such work certainly exceeds the possibilities of a teacher, alone in her/his class.

In the context of ICT distance learning, some experimentations (Joab & al 2003) allowed teachers to work collaboratively and gave birth to a new type of pedagogical resource, including a *scenario in use* (Allen & al 1994, 1996) taking the management of artifacts into account. This seems to be a way to make orchestrations explicit (see Conclusion).

APPENDIX

Appendix 8-1
About practicals
(Trouche 1998, p.16)

Work method proposed to students

You have to study this problem as a mathematical researcher. The most important thing therefore is the writing of a research report, in which you will note your approaches (even dead-ends), methods, tools used (if you use a calculator, you will specify which applications, which gestures, etc.).

If you solve the problem, so much the better! If not, the research done will not be useless! One learns from failures as well as from successes. What will be assessed will be more the relevance of your methods than the results themselves.

Your research report obviously has to be readable. But don't think of it as a final examination: it will necessarily bear marks of hesitation linked to each research process.

Working in pairs calls for collaboration, putting ideas in common and sharing tasks: a team enterprise. Avoid becoming too specialized in function (for example: always the same student using the calculator, always the same writing the research report): prefer task rotation!

During practicals, you have to study the following *main question*. You will probably have no time to study extra questions: you will tackle this problem later, in order to go further than the main question.

A example of practical text

Main question

How many figures 0 are there at the end of the numerical value of these expressions: 10!, 100!, 1000!, 1997!

Extra question

n being a given positive integer, a sequence u(n) is defined by "the number of figures 0 at the end of n!". Can you define this sequence for your calculator (and so answer easily the main question above)?

NOTES

1. It has been difficult to find good labels for each work method. Other difficulties appear when translating them. Firstly we chose the five French labels *théorique*, *rationnel*, *scolaire*, *bricoleur* and *expérimentateur*. The label *scolaire*, criticized, has been replaced by *automate* in our most recent papers (Trouche 2004). At first, our English labels were *theoretical*, *rational*, *random*, *mechanical* and *resourceful* (Guin & Trouche 1999b). The new labels chosen here seem to us better adapted, even if a single expression cannot summarize the whole description itself: for example, the fourth work method cannot be only characterized by the fact the work is mainly restricted to the calculator.
2. In this paragraph, the word *class* means basic schooling structure. Our propositions are based on experimentations carried out in 10th, 11th and 12th grade classes.
3. What is true for instrument use is also true, more generally, for mathematical practice. Brousseau (1997) writes: "... doing mathematics is first, for a child, a social activity, not only an individual activity".
4. The notion of *frame*, in this sense, was introduced by Douady (1986, p.11): it is "made of the objects of a branch of mathematics, the relationships between these objects, their eventually diverse formulations and the mental images associated with these objects and relationships".
5. These are theoretical proposals, not yet implemented on a calculator and not experimented.
6. With the same goal, Texas Instrument has developed a *Symbolic Math Guide* for its symbolic calculators "to help students learn algebra and some aspects of calculus by guiding them as they develop correct text-book-like solutions. SMG can be used when a student first learns a topic or as quick review" (http://education.ti.com).
7. Some constraints of this artifact can be analyzed:
   - the connection with a calculator requires a special plug on it, available only on some calculator types;
   - the cable linking this artifact to a calculator is only 2 m long.
   The consequence of these two constraints is that this device is probably designed for the teacher's use. Bernard & al (1996) showed indeed that it is, when available in a classroom, connected to the teacher's calculator.
8. On the one hand, the word sherpa refers to the person who guides and who carries the load during expeditions in the Himalaya, and on the other hand, to diplomats who prepare international conferences.
9. This advantage is not a minor one. Teachers, in complex technological environments, are strongly prone to perform alone all mathematical and technical tasks linked to the problem solving in the class (Bernard & al 1996).
10. Healy (2002) identified a major difference between instructional theories drawing from constructivist perspectives and those guided by sociocultural ideologies, which related to the primacy assigned to the individual or the cultural in the learning process. Constructivist approaches emphasise a *filling-outwards* (FO) flow in which personal understandings are moved gradually towards institutionalized knowledge. A reverse *filling-inwards* (FI) flow of instruction described in sociocultural accounts stresses moving from institutionalized knowledge to connect with learners understandings. Teaching interventions in Healy's study were therefore designed to allow investigation of these two different instructional approaches: the FO approach aimed to encourage the development

of general mathematical models from learners' activities; and the FI approach intended to support learners in appropriating general mathematical models previously introduced.

11. Meinadier (1991) thus estimates the size of the "task unit" in the context of computer use.

12. Vasquez Bronfman (2000, p.227) defined the *reflexive practicum*, an arrangement of quite the same nature, as "a frame, a way, aiming helping learners to acquire the art of working in uncertain (or undetermined) domains of their practice".

REFERENCES

Allen R., Cederberg J. & Wallace M. (1994) Teachers Empowering Teachers: A Professional Enhancement Model, *NCTM Yearbook: Professional Development for Teachers of Mathematics, National Council of Teachers of Mathematics*. Reston, VA.

Allen R., Wallace M., Cederberg J. & Pearson D. (1996) Teachers Empowering Teachers: Vertically-Integrated, Inquiry-Based Geometry in School Classrooms, *Mathematics Teacher* 90(3), 254-255.

Basque J. & Doré S. (1998) Le concept d'environnement d'apprentissage informatisé, *Journal of Distance Education/Revue de l'Enseignement à Distance* 13(1), 1-20.

Bernard R., Faure C., Noguès M. & Trouche L. (1996) L'intégration des outils de calcul dans la formation initiale des maîtres, *Rapport de recherche IUFM-MAFPEN*. Montpellier: IREM, Université Montpellier II.

Bernard R., Faure C., Noguès M., Nouazé Y. & Trouche L. (1998) *Pour une prise en compte des calculatrices symboliques en lycée*. Montpellier: IREM, Université Montpellier II.

Brousseau G. (1997) *Theory of Didactical Situations in Mathematics*, in N. Balacheff, R. Sutherland & V. Warfield (Eds.). Dordrecht: Kluwer Academic Publishers.

Chevallard Y. (1992) Intégration et viabilité des objets informatiques, in B. Cornu (Ed.), *L'ordinateur pour enseigner les mathématiques* (pp. 183-203). Paris: PUF.

Defouad B. (2000) *Etude de genèse intrumentale liée à l'utilisation de calculatrices symboliques en classe de Première S*, thèse de Doctorat. Paris: Université Paris VII.

Dehaene S. (1997) *La bosse des maths*, Editions Odile Jacob.

Delozanne E. (1994) Un projet pluridisciplinaire: ELISE un logiciel pour donner des leçons de méthodes, *Recherches en Didactique des Mathématiques* 14(1/2), 211-249.

Dorfler W. (1993) Computer use and views of the mind, in C. Keitel & K. Ruthven (Eds.), *Learning From Computers: Mathematics Education and Technology* (pp. 159-186). Berlin: Springer-Verlag.

Douady R. (1986) Jeux de cadres et dialectique outil/objet, *Recherches en Didactique des Mathématiques* 7(2), 5-32.

Faure C. & Goarin M. (2001) *Rapport d'enquête sur l'intégration des technologies nouvelles dans l'enseignement des mathématiques en lycée*. Montpellier: IREM, Université Montpellier II.

Guin D. & Trouche L. (1999a) Environnement "Calculatrices symboliques": nécessité d'une socialisation des processus d'instrumentation, évolution des comportements d'élèves au cours de ces processus, in D. Guin (Ed.), *Calculatrices symboliques et géométriques dans l'enseignement des mathématiques, Actes du colloque francophone européen* (pp. 61-78). Montpellier: IREM, Université Montpellier II.

Guin D. & Trouche L. (1999b) The complex process of converting tools into mathematical instruments: the case of calculators, *International Journal of Computers for Mathematical Learning* 3, 195-227.

Guin D. & Trouche L. (2002) Mastering by the teacher of the instrumental genesis in CAS environments: necessity of instrumental orchestration, in E. Schneider (Ed.), *Zentralblatt für Didaktik der Mathematik* 34(5), 204-211.

Healy L. (2002) *Iterative design and Comparison of Learning Systems for Reflection in Two Dimensions*, Unpublished PhD. London: University of London.

Hershkowitz R. & Kieran C. (2001) Algorithmic and meaningful ways of joining together representatived within same mathematical activity: an experience with graphing calculators, in M. Van den Heuvel-Panhuizen (Ed.), *Proceedings of PME 25* (Vol. 1, pp. 96-107). Utrecht: Freudenthal Institut.

Houdé O., Mazoyer B. & Tzourio-Mazoyer N. (2002) *Cerveau et Psychologie*. Paris: PUF.

Joab M., Guin D. & Trouche L. (2003) Conception et réalisation de ressources pédagogiques vivantes, des ressources intégrant les TICE en mathématiques, in C. Desmoulins, P. Marquet & D. Bouhineau (Eds.), *Actes de la conférence EIAH 2003* (pp. 259-270). Paris: ATIEF & INRP.

Meinadier J.P. (1991) *L'interface utilisateur*. Paris: Dunod.

Monaghan J. (1997) Teaching and learning in a computer algebra environment: Some issues relevant to sixth-form teachers in the 1990's, *The International Journal of Computer Algebra in Mathematics Education* 4(3), 207-220.

Mouradi M. & Zaki M. (2001) Résolution d'un problème d'analyse à l'aide d'un logiciel de calcul formel, *Recherches en Didactique des Mathématiques* 21(3), 355-392.

Noss R. & Hoyles C. (1996) Windows on Mathematical Meanings, *Learning Cultures and Computers* (pp. 153-166). Dordrecht: Kluwer Academic Publishers.

Pitrat J. (1990) *Métaconnaissance, futur de l'intelligence artificielle*. Paris: Hermès.

Proust C. (2000) La multiplication babylonienne: la part non écrite du calcul, *Revue d'histoire des mathématiques* 6, 293-303.

Rabardel P. (2000) Eléments pour une approche instrumentale en didactique des mathématiques, in M. Bailleul (Ed.), *Les instruments dans la pratique et l'enseignement des mathématiques, Actes de l'école d'été de didactique des mathématiques* (pp. 202-213). Caen: IUFM.

Rabardel P. (2001) Instrument mediated activity in Situations, in A. Blandford, J. Vanderdonckt & P. Gray (Eds.), *People and Computers XV - Interactions Without Frontiers* (pp. 17-30). Berlin: Springer-Verlag.

Rabardel P. & Samurçay R. (2001) From Artifact to Instrumented-Mediated Learning, *New challenges to research on learning*, International symposium organized by the Center for Activity Theory and Developmental Work Research, University of Helsinki, March, 21-23.

Robert A. & Robinet J. (1996) Pour une prise en compte du méta en didactique des mathématiques, *Recherches en Didactique des Mathématiques* 16(2), 145-176.

Ruthven K. (1997) Computer Algebra Systems (CAS) in advanced-level-mathematics, *Report to SCAA*. Cambridge: School of Education, University of Cambridge.

Ruthven K. (2002) Instrumenting Mathematical Activity: Reflections on Key Studies of the Educational Use of Computer Algebra Systems, *The International Journal of Computers for Mathematical Learning* 7(3), 275-291.

Trouche L. (1996) *Enseigner les mathématiques en terminale scientifique avec des calculatrices graphiques et formelles*. Montpellier: IREM, Université Montpellier II.

Trouche L. (1998) *Expérimenter et prouver: faire des mathématiques au lycée avec des calculatrices symboliques*. Montpellier: IREM, Université Montpellier II.

Trouche L. (2000) La parabole du gaucher et de la casserole à bec verseur: étude des processus d'apprentissage dans un environnement de calculatrices symboliques, *Educational Studies in Mathematics* 41(2), 239-264.

Trouche L. (2001) Description, prescription, à propos de limite de functions, in J.B. Lagrange & D. Lenne (Eds.), *Calcul formel et apprentissage des mathématiques, Actes des journées d'étude Environnements informatiques de calcul symbolique et apprentissage des mathématiques* (pp. 9-26). Paris: INRP.

Trouche L. (2003) From Artifact to Instrument: Mathematics Teaching Mediated by Symbolic Calculators, *Interacting with Computers* 15(6), 783-800.

Trouche L. (to be published in 2004) Managing Complexity of Human/Machine Interaction in Computerized Learning Environments: Guiding Student's Command Process Through Instrumental Orchestrations, *The International Journal of Computers for Mathematical Learning*.

Vasquez Bronfman S. (2000) Le practicum réflexif: un cadre pour l'apprentissage de savoir-faire. Le cas du campus virtuel des nouvelles technologies éducatives, *Sciences et techniques éducatives* 7(1), 227-243.

Wartofsky M. (1983) From genetic epistemology to historical epistemology: Kant, Marx and Piaget, in L.S. Liben (Ed.), *Piaget and the foundations of knowledges*. Hillsdale, N.J., Lawrence Erlbaum.

Chapter 9

THE INTEGRATION OF SYMBOLIC CALCULATORS INTO SECONDARY EDUCATION: SOME LESSONS FROM DIDACTICAL ENGINEERING

Michèle Artigue

Equipe DIDIREM & IREM

Université Paris VII Denis Diderot, France

artigue@math.jussieu.fr

Abstract: In this chapter, we approach the question of the integration of symbolic calculators into education, through the analysis of two didactic engineering projects developed and experimented by the research teams ERES (Université Montpellier II) and DIDIREM (Université Paris VII). The first project concerns exact and approximate computation, and the equivalence of algebraic expressions; it was planned for grade 10 students (15-16 year-old students); the second project concerns the teaching of the derivative to grade 11 scientific students. Through retrospective analysis of these two experiments, and by using the theoretical frames and approaches developed in the previous chapters, we investigate the problems raised by the integration of symbolic calculators into secondary mathematics education. This leads us, in the last part of this chapter, to discuss the viability of such an integration.

Key words: Algebra, Derivatives, Didactic engineering, Instrumental genesis, Viability conditions.

1. INTRODUCTION

The evolution of the mathematical field has always been dependent on the computation tools available and, as was shown in Chapters 2 and 3, the development of software for symbolic computation has had an increasing influence on mathematical practices and even problematics[1]. School, as is the case every time that it faces an evolution of scientific and/or social practices, can neither stand apart from this evolution, nor ignore the new needs it generates. But, as was underlined in Chapter 1, the appropriate form of adaptation by School is not obvious and is the source of many tensions. These tensions become easily understandable if one considers the reasons which, in the culture, legitimize mathematics education. What School wants firstly to transmit, through mathematics education, is some *mathematical culture* and the values of this culture: a way of approaching the world and a way of arguing which seem to characterize this discipline[2]. As evidenced by research in the philosophy and epistemology of mathematics, such characterization is more difficult than it appears at first sight, but beliefs about mathematics are well anchored in the culture and they *condition the relationships* that School develops with computational tools.

Professional worlds as well as society at large have a pragmatic relationship with computational tools: their legitimacy is mainly linked to their efficiency. But what School aims for, even in the most professional streams, is much more than developing an effective instrumented mathematical practice. The educational legitimacy of tools for mathematical work has thus both epistemic and pragmatic sources: *tools must be helpful for producing results but their use must also support and promote mathematical learning and understanding.* Answering this question is not simple. Mathematical needs are not absolute, independent from the evolution of problematics, of scientific and social practices; and the evolution of these is itself dependent on technological evolution. Up to now, nevertheless, school mathematical values and needs have tended to be seen as something rather absolute and independent from technological evolution. This position is less and less sustainable and *new balances need to be found.*

All technologies are not equally affected, but software for symbolic computation (CAS) is especially sensitive to these issues of educational legitimacy, all the more as what is at stake in secondary education is not the learning of symbolic computation as a specific domain, but the use of symbolic tools in mathematical education. Any instrumentation of these tools certainly requires some awareness of the problems and issues attached to formal calculus, as well as those arising from the representation of numbers and algebraic expressions, and from the validity, effectiveness and complexity of algorithms, but this awareness can remain very superficial. The *dominant logic is that of the tool* and this is often the only one visible.

While obeying the same logic, the use of dynamic geometry systems (DGS) does not raise the same problems. They were *conceived as educational tools* and are used in a domain -- geometry -- which crystallizes for our culture the values of mathematical reasoning and proof. It seems evident that DGS cannot take charge of all the mathematical work expected from the student in geometry. The situation is quite different with CAS. Our culture is little sensitive to the richness of the modes of reasoning which underlie numerical and algebraic computations, and to the learning of these modes, and it tends to reduce computation to the execution of algorithms. With symbolic software efficiently taking over this execution, teaching values are strongly destabilized and the integration of such tools into secondary education generates unavoidable tensions.

And even if one takes account of the evolution of practices and mathematical needs, and declares an ambition to train students to work, think, reason, produce, in a world instrumented by these technologies -- as the current world is -- the problem is still not solved. What remains open is the *question of the strategies* to be implemented to reach such an objective. As shown in previous chapters, such strategies have to *manage jointly and coherently* the development of both mathematical and instrumental knowledge, and this is far from being an easy task.

The research underlying this book takes this perspective and therefore can only have, with regard to the current state of integration, a prospective value. Even if it seems reasonable to think that, in the near future, symbolic calculators will be as common at secondary level as graphic calculators are now, today these technologies remain more or less marginal in secondary education. We are now in a phase where, while being in principle legitimate, they are not a living part of the educational institution. Thus, to carry out the experiments presented in this book, it was necessary to create protected environments allowing the researchers to study didactic processes which the usual environment did not allow them to observe. Thus the results which arise from these experiments have obvious limitations but, as we will try to show, they are nevertheless illuminating.

In this chapter, revisiting some engineering work which has been developed, we would like to give a synthetic view of the main contributions that this work offers for reflection on the integration of ICT into secondary mathematics education. These contributions are diverse. They are firstly of a *theoretical nature* as they provide us with a framework for approaching learning and teaching processes in symbolic environments, and with ideas allowing us to make this framework operational, such as those attached to instrumentation processes, both in their personal and institutional dimensions. This frame is more generally situated within the anthropological approach developed by Chevallard (Box 5-1). Then they are of an *experimental nature*, bringing us detailed information about the way in which students can jointly develop their mathematical and instrumental knowledge, about the new needs that these elaborations require, and the ways in which teachers can take charge of them efficiently. All these contributions have been presented and discussed in the previous chapters. In this chapter, we would like to draw on them to analyze some of the engineering strategies which were developed, to analyze and discuss the choices underlying these, to point out some interesting regularities, and finally to question their *conditions of viability*. Even if they were carried out in privileged environments, these engineering strategies could not wholly escape institutional and cultural constraints; their actors faced problems of compatibility with standard mathematics education, and many other problems. The ideas that they offer are essential in order to think about less marginal uses, to think also about one key issue, that of teacher training.

For obvious reasons of space, we could not seriously analyze in this chapter the whole engineering work developed by the DIDIREM and ERES teams, not to mention others. Indeed, this work covers the teaching of algebra and calculus over the three years of senior high school (from grade 10 to grade 12). We have preferred to focus on some examples and to evidence, through these examples, the type of work carried out, the problems met, the results obtained. We have chosen two examples. The first example deals with numbers and algebra in grade 10. It corresponds to a piece of 'local' engineering (over three sessions). The scenario and the reported observations correspond to a first experiment, carried out with students who, at the same time, were discovering the TI-92 calculator. The second example concerns the teaching of the derivative to grade 11 scientific students. This is a more 'global' product. The analysis, furthermore, takes into account the evolution of the design between the first and the second year of experiment. These two examples thus present different characteristics and offer complementary insights for reflection on the questions raised in this introduction.

2. EXACT AND APPROXIMATE COMPUTATIONS, EQUIVALENCE OF EXPRESSIONS (GRADE 10)

The ERES team in Montpellier decided to link familiarization with the TI-92 to some mathematical work on exact and approximate computations, and on the *equivalence of expressions* (Guin & Delgoulet 1997). Before going into the detail of this engineering and into its analysis, we shall clarify what it seems in principle possible to expect from this kind of design, considering it as a particular but representative example of a wider set.

2.1 The mathematical and instrumental potentialities a priori of such an engineering

2.1.1 Rethinking the relationships between exact and approximate computations

The use of numeric and graphic calculators, in secondary education, tends to separate two worlds: that of exact computations, done by hand, and that of approximate computations performed by the calculator, and used when exact computations are impossible or too complex. In this practice of approximate computation, the quality of the estimates obtained is not seriously evaluated, as was shown by Birebent's recent thesis (Birebent 2001)[3]. This vision of the relationships between exact and approximate computations is well anchored in the culture but does not reflect current scientific functioning[4].

With symbolic calculators, both types of computation become instrumented by the machine. We can imagine that this change helps to approach the question of relationships between exact and approximate computations in a more adequate way and can have positive effects on the students' understanding of approximation.

2.1.2 Working on the relationships between sets of numbers

It is well known that the use of scientific and graphic calculators tends to obscure the distinctions between numbers. Real numbers, whether they be decimal, rational or irrational, uniformly appear in the form of a decimal estimate limited to at most a dozen places.

This allows the true status of the numbers involved in mathematical work to remain fuzzy and does not help students to distinguish between numbers and their representations. The secondary curriculum, at least in France, does not give students the means to clarify the situation, as was evidenced by Bronner (1997), and students' conceptions of numbers continue to be fuzzy beyond the secondary level. Bronner shows, for instance, that even students preparing for the CAPES[5] examination can have difficulties with questions such as the following: what is the exact nature of the number $\sqrt{13.21}$? What do you think about the decimal value given by the calculator?

By offering the possibility of working both on decimal estimates and on symbolic expressions of rational and irrational numbers, the symbolic calculator changes the conditions of mathematical work and allows the teacher to introduce issues that traditional education has difficulty in bringing to life.

2.1.3 Working on the sense of algebraic expressions

Considering algebraic objects, it is interesting to distinguish, according to Frege, their *sense* and their *denotation*[6]. In the algebraic work, transformations carried out (for instance in solving an equation or an inequation) must preserve the denotation of the objects involved but, as soon as one goes beyond routine tasks, transformations are driven by the sense of the expressions involved. The different equivalent forms that an algebraic expression can take do not give us the same information about this expression and do not present the same utility according to the problem to solve. The expanded form $x^4 - 9x^3 - 5x^2 - 45x - 50$ is useful if one wants to study the behavior of this polynomial when x tends towards infinity but the factorized form $(x + 1)(x - 10)(x^2 + 5)$ is better adapted if one wants to find the roots of this polynomial or to study the sign of this expression according to the values of x (to study for example the variations of a function whose derivative is this polynomial modulo a factor of constant sign). In France, at junior high school level, students are not given real autonomy in planning computations according to the sense of the manipulated objects. It is when students enter senior high school that developing such an *intelligence of algebraic computation* becomes a real concern of mathematics education.

The use of symbolic calculators can be particularly useful to support this learning. In a paper-and-pencil environment indeed, limitations in the students' abilities make it necessary to severely limit the complexity and variety of the expressions manipulated. These limitations tend to reduce algebraic computation to a small number of routines. Thus, the intelligence of computation does not have an appropriate space to develop. Working with symbolic calculators modifies the economy of the didactical system. On the one hand, it becomes less expensive to work with more complex expressions if the transformations are made by the machine, and the necessity of choosing appropriate commands makes explicit and central the planning of computations according to their precise aims. On the other hand, the machine does not always work as expected. The simplifications which it automatically performs on the expressions entered, the form in which it expresses results, are often different from those usual in the paper-and-pencil environment at school. Thus, students are necessarily faced with problems of equivalence and have to develop competences for dealing with these efficiently, both mentally and with the help of the calculator. Another point is that transformations which seem elementary can be impossible to obtain, for example in trigonometric computations, while the machine gives immediate results for computations which would be very painful, even impossible to perform by hand. This faces the student with a rich and unexpected algebraic landscape. Learning to understand it in order to become an efficient user of the machine is a new concern for algebraic education which can help to develop what we called above the intelligence of algebraic computation.

2.1.4 Making students sensitive to key issues in the field of symbolic computation such as simplification and data representation

The use of symbolic calculators, even if these are mainly considered as tools for mathematical activity, faces students and teachers with key issues of symbolic computation seen as a discipline, such as *simplification* and representation issues. Beyond the tool dimension of CAS, what is at stake here is their object dimension, according to the distinction made by Douady between these two facets of mathematical concepts (Box 9-1).

Box 9-1.

Tool and object dimensions of a mathematical concept
(Douady 1986, p.9)

An important part of the activity of mathematicians consists of setting up and solving problems. To do that, researchers are led to create conceptual *tools*, to which have been added technical tools (such as computers and software today). Due to the needs of communication inside the scientific community, the concepts thus created are de-contextualized, and expressed in the most general form possible. They integrate, then, the body of already built knowledge, extending it or substituting for some pieces of it. In this process, they gain an *object* status. It also may happen that researchers directly create new objects in order to reorganize a branch of mathematics.
For these reasons, we say that a concept is a *tool* when we focus our interest on the use made of it for solving some problem. By *object*, we understand the cultural object having some place in a larger body, that of scholarly knowledge at a given moment, socially acknowledged.
Knowing mathematics is thus two-fold: on the one hand, it means having some functional availability of certain notions and theorems for solving problems, for interpreting these and rising new questions. On the other hand, knowing mathematics means also identifying notions and theorems as elements of a corpus, scientifically and socially acknowledged. This is also a matter of articulating definitions, formulating theorems and proving these. In this case, notions and theorems have the status of object.
Summarizing, a concept is a tool when it is used for solving problems internal or external to the mathematical field; it is an object when it is worked on for itself or for its relationships with other objects.

The entry of an expression into a symbolic calculator automatically produces an evaluation and simplification of this expression as shown by the screen reproduced in Box 9-2. The entered expression appears on the left side of the screen, and one can thus check that it was correctly entered; the simplified form appears on the right side. Sometimes these two forms are identical or differ only in the order of their terms, sometimes changes are more substantial: a polynomial expression can be factorized or expanded, a rational fraction can be reduced to simple elements, quotients can be simplified by factors, radicals can disappear... Why such transformations? Are they always legitimate? Do they always correspond to simplifications? And what does simplifying really mean? Here, there is in principle a fantastic opportunity for approaching issues that are generally only a matter of didactic contract in school algebra (Box 8-6).

Box 9-2.

Different effects of the key ENTER

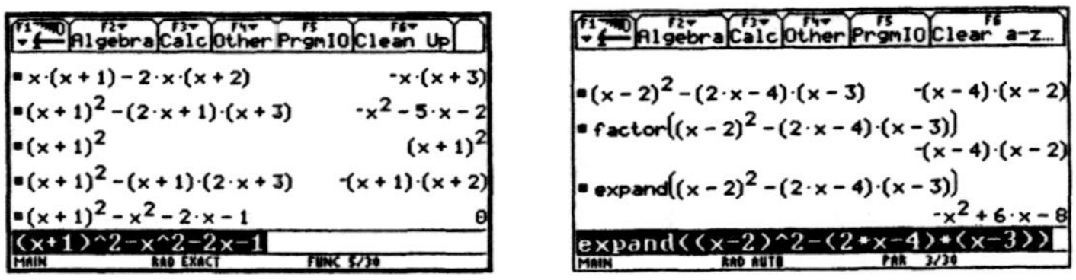

Let us consider simplification. On symbolic calculators, such as the TI-92 and TI-89, it is automatically performed; with software such as MAPLE or MATHEMATICA, it is no longer automatic, and different options are offered[7]. But, whatever be the case, what makes one expression simpler than another? If one considers as a criterion the number of operations to be performed to calculate the value of a polynomial for a given value of the variable, the Horner's form[8] is doubtless simpler than the expanded one. If one has to find its roots, a factorized form will be simpler. And this is not even systematic: to determine the roots of $x^6 - 1$, this expanded form is more illuminating than the rational factorization $(x - 1)(x + 1)(x^2 + x + 1)(x^2 - x + 1)$ given by the machine. Thus, simplification cannot be defined in an absolute way, and those produced by symbolic software are not necessarily the best adapted to the user's aims. Working with them provides students with a clear opportunity to understand this essential fact, and beyond that, to learn to combine simplification commands with other software commands or with paper-and-pencil work in order to achieve their mathematical goals.

Checking the equivalence of expressions is a problem whose solution relies on the possibility of *recognizing* a mathematical object through the various representations which can be associated with it. Here, the notions of *canonical form* and *normal form* play a fundamental role (Chapter 2, § 1.3). In the best cases, if E is a class of symbolic expressions, there is a calculable mapping φ from E to E which is such that if p and q represent the same object, they have the same image under φ: the canonical representation of this object. If such a mapping is implemented in a piece of software, this allows it to check equivalence for expressions from the class. Unfortunately, such canonical forms only exist for a few classes of objects. A weaker but very useful property is the existence of a mapping φ which verifies the condition given above only for the subclass of the objects equivalent to 0. Once again this makes possible a test that p and q represent the same object, by verifying this time that p – q is in the subclass of 0. This also makes it possible to check automatically that the results of simplifications do not involve factors from this subclass. In fact, for a given piece of software, only particular classes of symbolic expressions admit canonical or normal forms. This explains why equivalent forms are not necessarily recognized or why, in some cases, while the software does not return a positive answer to the test p = q, it does to the test p – q = 0.

An *efficient instrumentation* requires some sensitivity to these problems which are linked to the representation of formal objects, and to the development of instrumented schemes allowing the user to take these phenomena into account in computations.

Here, we approached the question of *data representation* through simplification and equivalence, referring to the particular engineering which we are analyzing. This question could be also treated in a more direct way, by trying to reach the internal representation of expressions, through the specific commands offered by symbolic software for this purpose. MAPLE, for example, via the *Tree* command, gives access to the tree associated with a given symbolic expression. Texas calculators have no equivalent command, but the TI-92 Plus and the TI-89 include a *Part* command, which makes it possible to obtain a progressive decomposition into a tree (with the limitation that any node is at most binary). This can provide an interesting alternative approach for secondary students, for instance grade 10 students, since associating a tree with a symbolic expression is not obvious at this level. Such a syntactic analysis assisted by the machine can help the students to differentiate expressions which they tend to confuse.

We thus see that this type of engineering has, in principle, many interesting potentialities. What precise choices were made in the particular piece of engineering carried out by the ERES team? What were the results of these? This is what we shall now examine.

2.2 The engineering

This piece of engineering consists of three laboratory sessions which are also the first three sessions with the TI-92 for the students. This inevitably conditions the choices made by the ERES team as the teacher will be obliged to spent a substantial amount of time on the collective appropriation of the machine during these sessions.

Thus session 1 begins with a presentation of the machine using overhead projection. Documents concerning the keyboard, the screen, the status line and the different options for computation offered by the menu *MODE* are also given to the students. The machine is then configured in exact mode and students are shown how to obtain approximate results with the *Diamond key*, without changing this mode (approximate detour, Chapter 6, § 2.3.1). The teacher also asks them to systematically note error messages in the individual work which they will do, as well as the computation which produced the message, and how they solved the problem.

In session 2, the use of menus is approached more systematically, with particular attention to the menus *MATH, CATALOG, CHAR* and the different menus offered by the application *HOME*. The teacher shows shortcuts which save time, and how to copy an expression from the screen into the command line. Finally, the syntax of the main commands to be used in this session, *Factor* and *Expand,* is introduced. A document concerning the menus of the application *HOME* is also distributed.

2.2.1 The first session

A priori analysis

This session (Box 9-3), only involves numerical computations. Three series of computations are proposed to the students, including respectively decimal numbers and fractions (series 1), powers (series 2), radicals (series 3).

Box 9-3.

TP1

(Guin & Delgoulet 1997, p.42)

	Series 1: with decimal numbers and fractions	Normal Float		Scientific Fix 5
		Results with ENTER	Results with ◆ ENTER	Results with ◆ ENTER
1	$21.753\times(0.123-3.5426)$			
2	0.1234567×136963			
3	$30+\dfrac{3}{3+\dfrac{3}{5}}$			
4	$\dfrac{864}{10-\dfrac{75}{35+3}}+\dfrac{84-\dfrac{1}{2}}{35-\dfrac{3}{55}}$			

	Series 2: with powers	Normal Float		Scientific Fix 5
		Results with ENTER	Results with ◆ ENTER	Results with ◆ ENTER
5	$1.725\times10^{-4}+3.02\times10^{-5}-0.04\times10^{-3}$			
6	$\frac{5^{23}}{2^{15}}\times\frac{3^{16}}{5^{14}}\times\frac{2^{17}}{3^{5}}$			
7	$\frac{10^{23}}{23^{10}}$			

	Series 3: with square roots	Results with ENTER
8	$\sqrt{75}$	
9	$3\sqrt{18}-\sqrt{128}+10\sqrt{32}$	
10	$(\sqrt{3}-\sqrt{6})^2$	
11	$(\sqrt{3}-\sqrt{5})^2$	
12	$\frac{3}{\sqrt{3}-\sqrt{5}}+\frac{2}{2+\sqrt{5}}$	
13	$\frac{\sqrt{5}-\sqrt{3}}{\sqrt{3}-5}+\frac{2}{\sqrt{3}+\sqrt{5}}$	
14	$\frac{\sqrt{5}-3}{\sqrt{3}-5}+\frac{2}{\sqrt{3}+\sqrt{5}}$	

For the series 1 and 2, an exact computation and parallel approximate computations, the first in floating mode, and the second in scientific mode with the number of decimal places fixed at 5, are asked for. For the series 3, only an exact computation is asked for. The essential aim of this activity is to understand and to interpret the results given by the machine. The students have to note the results they obtain, to *interpret* them (they are told that results may be different from those expected), and to compare the results of apparently similar computations (expressions 10 and 11 for instance).

The first two series also have an aim of establishing some link with previous activities that these students have carried out with scientific calculators. Computations 1, 2, 5, 6 and 7 have already been proposed in a previous session. Expression 2, with a scientific calculator, leads to the result 16909 but elementary reasoning about the last digits of the numbers involved shows that this result cannot be exact. The TI-92 where results can be expressed with 12 digits, gives the exact result in the normal floating mode, 16909.0000021, and, naturally, in the exact mode, this time with a fractional representation, $\frac{169090000021}{10000000}$. The fractional expressions 3 and especially 4 are more complex than those ordinarily used in a paper-and-pencil environment. While showing that the symbolic calculator immediately turns these expressions into their canonical fractional representation, these examples can serve to verify students' ability to deal with brackets, and to make them aware of the fact that they can check their entries by looking at the left side of the screen. The result of computation 6 involving powers can be, for its part, easily predicted or at least interpreted after the event.

With the series 3, students enter a new domain: that of exact computation with expressions involving radicals. It is no longer the comparison between exact and approximate results which is at stake but some understanding of the way in which such exact computations are performed by the machine. Expressions 8 and 9 show that expressions of the type $a\sqrt{b}$ are simplified by extraction of the squared factors of b. Expressions 10 and 11 evidence a more complex functioning of simplification: expression 10, where a factorization is possible, is factorized into $3(\sqrt{2}-1)^2$ whereas expression 11 is expanded, $-2\sqrt{15}+8$. Note however that if one enters the expression $3(\sqrt{2}-1)^2$, it is automatically expanded, giving $-3(2\sqrt{2}-3)$ with a factor -3 (while in paper-and-pencil work, the factor would rather be 3), but this computation is not asked for. The TI-92 recognizes the equality between expression 10 and this last expression, as could be expected, the canonical form $a\sqrt{2}+b$ being accessible by simple expansion and extraction of squares.

Expressions 12, 13 and 14 can, in principle, be represented in a homogeneous way by a canonical expression of the type $a\sqrt{15}+b\sqrt{5}+c\sqrt{3}+d$, easy to obtain by multiplication by the conjugates of denominators and by reduction. Even so, this is not systematically the case: expression 12 is transformed into $\frac{\sqrt{5}}{2}-3\frac{\sqrt{3}}{2}-4$, whereas, for expression 13, a factorization is made leading to the expression $\frac{(\sqrt{5}-\sqrt{3})(\sqrt{3}-17)}{22}$ and, for expression 14, one gets a mixed representation, $-\frac{(\sqrt{5}-3)(\sqrt{3}+5)}{22}+\sqrt{5}-\sqrt{3}$. Let us note also the presence, which can be surprising for a student, of the sign $-$ in front of the factors in the expressions 13 and 14. Furthermore, the TI-92 recognizes the equality of the expression 12 and its canonical form but this is no longer the case for expression 13 (negative results are obtained both when testing equality and equality to 0 of the difference).

Working on these expressions, which remain relatively simple, demonstrates the differences between the treatment in a paper-and-pencil environment and that of the machine, even if both types of treatment lead to the elimination of radicals from denominators. The simplifications carried out by the machine depend strongly on the particular characteristics of the expressions. Furthermore, such work shows that the formal equality of two expressions can be recognized by the machine, even if it does not transform the first one into the second one. Nevertheless this is not systematic for expressions involving several radicals, even if these have a canonical representation, from a mathematical point of view.

In the *a priori* analysis that we have developed so far, we made use of our mathematical knowledge about computations involving radicals, and we performed paper-and-pencil and machine computations which were not explicitly asked for, so as to understand better the functioning of the calculator. We cannot expect the same behavior from grade 10 students whose familiarity with radicals is very limited and who have just begun to use the machine. How did they react to the proposed tasks and what was brought about by their interactions with the teacher in the summing up phase of the lesson?

A posteriori analysis

The report written by the ERES team on this experiment confirms the importance of the collective phase of appropriation. In the group work by students, with few exceptions, handling the machine did not raise problems. As could be expected, students received different error messages, many of these being linked to problems in the management of brackets, but they were able to overcome them. In spite of what had been pointed out to them in the collective phase, they did not spontaneously check their entries by looking at the left side of the screen, and the teacher was obliged to intervene in that respect. Furthermore, contrary to the expectations of the teacher, the students were not surprised at discovering the possibilities of exact computation with the machine and did not remember having already made some of the proposed computations, with the exception of expression 2. Computations with radicals gave rise to few interpretations; results were noted by the students but no more. Here we touch on the evident limits of this type of exercise: surprise effects and the resulting motivation for understanding can only exist if there is some *expectation*. The familiarity of these students with this type of computation was much too limited to induce spontaneous predictions on their part, in contrast to the case for us designers. Moreover, the work of forming such a prediction is only of reasonable cost if it can be partly done *mentally* and does not oblige the student to carry out very detailed calculations. This was not the case for the majority of these students, and they simply did what they were asked, fulfilling the minimum obligations of the didactic contract. For the expressions of the third series, prediction would have needed to be *collectively initiated.* In the summing up phase of the lesson, due to time constraints, the teacher could not go beyond the interpretation of the first four computations with radicals, leaving the last three, the richest, for the following session.

2.2.2 The second session

In the second session (Box 9-4), the collective appropriation of the machine continued as explained above and commands for the factorization and expansion of algebraic expressions were introduced. It was shown that the TI-92 permits computations generally regarded as forbidding to be done quickly, and factorization of expressions beyond what students' mathematical knowledge usually makes possible at this level of schooling. Taking into account the results of the first session, students were explicitly asked the following: when the machine produced an unexpected or incomprehensible result, they had to record the expression concerned, the nature of the algebraic manipulation involved: expansion or factorization, the result given by the TI-92, their remarks and comments, as well as their interpretation of the result, if any.

Box 9-4.

TP2

(Guin & Delgoulet 1997, p.55)

	$(2x-5)^2-(3x-4)^2$	$(\frac{1}{2}x-\frac{3}{4})^2+(2x-3)(x-1)$	x^2-5x+6
Factorization			
Expansion			
	x^2+3	x^2-5x-9	$(2x-3)^2$
Factorization			
Expansion			
	$(5x-3)^2(x-1)-4(x-1)$	$(2x+2\sqrt{2}-3)(2x-\sqrt{2}-3)$	$(x\sqrt{2}-\sqrt{3})(x\sqrt{3})-\sqrt{2}$
Factorization			
Expansion			

A priori analysis

In the first series, factorization precedes expansion. None of the expressions is factorized immediately but the first one is a difference of two squares and, in the second one, the factor $(2x-3)$ although masked by a fractional coefficient is common to both terms of the sum. Other expressions are, as regards factorization, outside the field of students' knowledge. The command *Factor* works with the first three expressions, and the command *Factor (, x)* is necessary for the last one, seemingly of the same type but with irrational roots[9]. The fourth expression is not factorizable in the reals and the machine does not transform it. The expansion of the factorized forms does not offer any surprises.

In the second series, expansion precedes factorization. Three of the four expressions are already factorized and, in the last one, a common factor $(x-1)$ is visible in both terms of the sum. The expansions do not raise problems. The radicals disappear in the third expression, $4x^2-12x+1$. Notice that, in the last expression, the product $\sqrt{2}\sqrt{3}$ is not transformed into $\sqrt{6}$ as would be done in a paper-and-pencil environment. As regards the factorizations of the expansions, with the command *Factor* one finds the original expression as expected. For the third one, one finds the initial expression only by using *Factor (, x)*. The last one is factorized through the command *Factor*, which is surprising as the roots are also irrational and the result is $\frac{\sqrt{6}(\sqrt{6}x-3)(\sqrt{6}x-2)}{6}$, which is not the initial expression. This time $\sqrt{6}$ is not decomposed into a product. As we can see, even if the expressions are not complex, here the student faces the fact that the same expression can admit various factorizations or expansions, that the machine privileges certain forms, but that it is not easy to anticipate what form exactly, in any particular case.

A posteriori analysis

The part of the lesson given over to the collective exploration of the different menus and the introduction of new commands took more than half an hour, then the students worked in groups on the worksheet. The teacher was obliged to intervene on different occasions as some students did not check their entries or did not take notes. Some differences also became visible between the students as regards the manipulation of the machine. Some felt very much at ease whereas others still needed to be helped.

As in the previous session, the teacher wanted to make the students aware of what he called *"contradictions or incoherences of the machine"* (in fact these reflect the complex problems raised by formal simplification as mentioned above, and the machine is not inconsistent, but when compared to the normalized forms of simplification employed in school algebra, its functioning can appear rather chaotic). He wanted to make them *sensitive* to the *differences* between machine and paper-and-pencil computations, but apparently the situation did not bring about this sensitivity, at least for the students concerned here, and the more precise instructions given, taking account of the previous session, were not effective.

In fact, the successful functioning of the situation would have required students to know how to *discriminate* between expanded, factorized and other equivalent forms. The implementation revealed that this was far from being the general case. No group except one, for example, noticed that the factorization given by the TI-92 for the last expression was different from the initial one, and the collective discussion showed that two groups at least, put out by the radicals, had not read the last two expressions given as factorized forms. Some students did not even link $\sqrt{2}\sqrt{3}$ and $\sqrt{6}$, which did not facilitate interpretation. In the face of such difficulties, students tended to convince themselves easily that the machine was inevitably right and passed on to the next task. Finally, the students' work with the machine, by facing them with more complex expressions than those they usually met, showed the *fragility* of the criteria they used for identifying factorizations and expansions, and the collective summing up provided a good opportunity for coming back to these questions.

2.2.3. The third session

The second session dealt only with polynomial expressions; in the third session (Box 9-5) fractional expressions appeared, and equivalence issues became explicit.

Box 9-5.

TP3

(Guin & Delgoulet 1997, p.65)

Series 1: Fill the table by using the corresponding commands					
	Expressions	ENTER	*Expand*	*Factor*	*ComDenom*
1	$\frac{3}{x-1}+\frac{5x-1}{x-2}$				
2	$\frac{x-4}{(x-2)(2x-1)}+2$				

Series 2: Which expressions in the first line are equal? Note your results in the last column. Do the same with the second line.					
	f(x)	g(x)	h(x)	k(x)	Results
3	$\frac{1}{x}+\frac{3x-5}{x-7}$	$\frac{3x^2-4x-7}{x(x-7)}$	$\frac{2}{x}+2+\frac{16}{x-7}$	$\frac{(x+1)(3x-7)}{x(x-7)}$	
4	$\frac{x^2-6x+2}{2x-1}$	$\frac{11}{4}-\frac{x}{2}+\frac{3}{4(2x-1)}$	$\frac{(x-\sqrt{7}-3)(x+\sqrt{7}-3)}{2x-1}$	$\frac{x}{2}+\frac{4-11x}{2(2x-1)}$	

Series 3: The expressions A and B are equal; how can A be turned into B or B into A with the TI-92. Is this always possible?		
	A	B
7	$6+\frac{35}{4(x-3)}-\frac{7}{4(x+1)}$	$\frac{6x^2-5x-4}{(x-3)(x+1)}$
8	$\frac{\sqrt{2}-1}{\sqrt{3}-2}+\frac{2}{\sqrt{3}-\sqrt{2}}$	$2+3\sqrt{3}-\sqrt{6}$
9	$\frac{2x^2-3x-2}{(x-4)(3x-1)}$	$\frac{(x-2)(2x+1)}{3x^2-13x+4}$

For this session, in order to focus classroom activity on interpretation and analysis, the worksheet was distributed to the students in advance. They had to do the computations at home and to note carefully the results and error messages. The plan was especially to discuss, during the collective summing up, the notion of equivalence and its checking both by paper-and-pencil techniques and with the machine.

A priori analysis

For the first series, two expressions are given and students have to note the results obtained by using the following commands: *Enter*, *Expand*, *Factor* and *ComDenom*. The machine automatically decomposes the first expression into simple elements, which also corresponds to the effect of *Expand*. The command *Factor* produces the reduction to the same denominator (in factorized form), the numerator being expanded (its roots are irrational). The *ComDenom* command gives numerator and denominator in expanded form. The second expression is left as such by *Enter*, is decomposed into simple elements by *Expand*, reduced to the same denominator and factorized by *Factor*, the *ComDenom* command having the same effect as on the first expression.

For the second series, the *ComDenom* command gives the expression $-g(x)$ with an expanded denominator, the command *Expand*: $\frac{16}{x-7}+\frac{1}{x}+3$, and the command *Factor*: k(x). For the second expression, *Expand* gives $-g(x)$, *Factor(, x)* gives h(x). Finally the second term of k(x) can be also obtained from $-g(x)$ by *ComDenom* (the denominator is then developed).

Thus passing effectively from one expression to the other with the machine supposes a preliminary analysis of the form of the targeted expression so as to judge which command(s) will make it possible to obtain the expression directly, if this is possible, or to get an expression close enough to make comparison easy. Of course the machine also offers the possibility of comparing expressions, by testing equality directly or by calculating differences. The equality f(x) = k(x) is returned as it was entered but the test $f(x)-k(x)=0$ gives the answer *true* and the computation of the difference gives 0.

The series 3 raises the question of transforming an expression to another equivalent one. The situation has apparently been designed so that this transformation is not always possible. As might be expected, for the first expression, B can be turned into A by *Expand* (if the order of the terms is not taken into account) but it is impossible conversely to turn A into B. However the numerator and the denominator can be obtained separately by applying to A the commands *GetNum* and *GetDenom*, two options of the submenu *Extract.* The third expression raises the same problems, in both directions this time. The second expression leads to equivalence problems similar to those met with radicals in the first session. If one enters the expression A, the form obtained is similar to that obtained for the expression 14 in session 1: $-(\sqrt{3}+2)(\sqrt{2}-1)+2\sqrt{3}+2\sqrt{2}$. If one tests directly the equality of both forms, the machine answer is *false*, but if one tests the equality to 0 of the difference, the answer becomes *true*.

As we can see, in this session also, the expressions, although relatively simple, allow for interesting work and one can *a priori* hypothesize that both the new organization chosen, and the increasing familiarity students have with the machine and this kind of algebraic task, are going to allow the teacher to engage students in analysis and interpretation, as wished. All the more so, as there is only one new command to learn: the command *ComDenom*.

A posteriori analysis

The collective discussion of the results obtained in the first series made it possible to clarify the functioning of the various commands. It was, for the teacher, the occasion to point out that understanding the effects of these commands was easier than understanding the effect of the automatic simplification activated by the *Enter* command.

For series 2, the students had difficulty in remembering the way they had verified the equalities. Apparently, many did not use the different possibilities that we outlined in the analysis *a priori*, and just tested the equality for a simple value of x. Two students checked it for $+\infty$ and $-\infty$, as if that ensured equality for every number. In the discussion, students suggested checking the equalities by systematically expanding the two expressions but only one of them suggested testing whether the difference is 0. Another suggested using the *Solve* command which is indeed effective for this type of expression. Once again, the work with the machine served as a *revelator of the fragility* of the students' mathematical knowledge, and evidenced up to what point mathematical knowledge was necessary to *efficiently pilot* the machine. This time, the session led to a making explicit of the methods used by the students to test the equality of two expressions, to the destabilization of some previously popular strategies, such as the test on one particular value, and to an enrichment of their instrumented techniques.

2.3 In conclusion

At the beginning of this part, we tried to elucidate the didactic potential of such algebraic work with a symbolic calculator. As is often the case, consideration of a particular engineering design in a particular context -- that of a class considered by the teacher as rather passive -- allows us to measure the *distance* between the *identification of potentialities a priori* and their *actual implementation* in a given class, even an experimental class as was the case here. In this engineering design, the authors relied on a lever which had been identified as a productive one in different research projects about DGS or CAS: the software was used as a provider of strange phenomena involving mathematics already known or to be learnt, and was cast as *means for understanding* these phenomena, for anticipating or even producing them, at will. Different examples have already been given in this book. Conditions necessary for the effectiveness of this lever, as pointed out by Laborde (1999), are the following: the phenomenon has to be easily identifiable; it has to arouse the students' curiosity enough for motivation towards understanding to be provided not only by the teacher; finally, the expected mathematical work has to be accessible to the students, with their mathematical and instrumental knowledge. These conditions are not easy to satisfy and, in particular, if the cognitive gap is too great, if a refuge is possible in tasks of simple computation or in poorly controlled trials, students' work risks being far from the expectations of the designers. Here, the students clearly had difficulty in developing expectations, and thus phenomena intended to be surprising to them were not so. In fact, the potential of this engineering lay more in the diagnosis which it allowed. By exposing the students to unusual expressions, the work with the machine provided evidence of a certain fragility of knowledge which could have remained invisible with simpler examples and more routine tasks. This led to a return, in this context of the instrumentation of a new object, to important questions in algebra and more extensive work on these. But, as evidenced by this experimentation, the success of such an enterprise requires a *precise piloting* by the teacher, extending beyond the careful choice of a progression in the proposed expressions and tasks and in the commands introduced and worked[10].

3. TEACHING THE DERIVATIVE TO SCIENTIFIC GRADE 11 STUDENTS

The second engineering design presented in this chapter deals with the teaching of the derivative to grade 11 students and was piloted by the research team DIDIREM (Artigue & al 1998, Artigue & Lagrange 1999). In the two classes where it was tested out, this piece of engineering followed some work on functional situations involving algebraic, numeric and graphic representations. This had accompanied students' familiarization with the TI-92. A first approach to the notion of limit had also been organized. As with the previous project, before going into more detail, we will clarify the interest we see in principle in this type of didactic construction. We shall emphasize especially the different categories of use of the calculator that such a project can involve, in the different phases of the teaching process.

3.1 Introduction

In the current French syllabus for senior high school, the teaching of elementary Analysis (or Calculus in the English-speaking world) is organized mainly around the notion of derivative and its applications. It is through this notion and the different perspectives which can be taken on it, that students begin to enter the interplay between *local and global points of view* on functional objects, which is crucial in this conceptual field, as underlined in the report on Computation produced by the already mentioned CREM (Chapter 3, § 1; Kahane 2002). As also pointed out in this report, a reasonably aim for Calculus teaching in high schools today cannot be an entrance into the field of formal analysis based on definitions in ε and ∂ but it can be a first approach which, while remaining more 'intuitive', prepares for the necessary reorganization of knowledge:

> Access to the field of Analysis and to the world of approximation is known to be difficult. The available time, the evident limits of the students' algebraic competence when this teaching begins, impose modesty. But if computation in Analysis wants to be more than a formal functional calculus, which seems desirable to us, it has to make students understand its essential values, through computations remaining at a reasonable level of complexity. From this point of view, we think that it is essential to develop two awarenesses as early as secondary education:
>
> - understanding that computation in Analysis differs from previous algebraic computation, in the interplay which it establishes between 'the local' and 'the global';
> - understanding that computation in Analysis involves, in a fundamental way, the notion of order of magnitude.
>
> It seems to us that some awareness can be reached with reasonable levels of formalization and technicality.

For instance, on entering the field of Analysis, linearity, previously perceived as a global phenomenon, has to take a local dimension and, as also underlined in the same report, while calculators can help this localization through the effects of the zooming they permit, nevertheless the essential work of mathematization remains:

> Today graphic calculators make it possible to illustrate local linearity very easily by means of successive zooms in the neighborhood of a point. They also allow its problematization. Indeed, if in the neighborhood of a point the graphic representation of a function tends to become a straight line, how can this phenomenon be characterized mathematically? Here, it is important to underline that visualization does not permit the order of the approximation to be checked; thus the mathematical object is not given but has to be built.

The engineering design on the derivative which we present and analyze below takes this perspective. It was designed in order to study how the use of tools of symbolic computation can support the realization of such ambitions and inform us about their *conditions of viability*.

Globally, the project was built around various phases which contribute to approaching the notion of derivative both in its *tool and object* dimensions (Box 9-1), in order to ensure a good *connection between the main points of view* related to the derivative at this level of education -- geometrical, numerical, kinematic and algebraic points of view -- and the associated reconstructions of knowledge, while developing naturally the technical competence necessary for work in this domain (Box 9-6).

In an engineering approach of this type, the TI-92 and the symbolic computations which it allows, can serve various purposes *a priori*. We clarify some of these below, and this will help us to situate the engineering design analyzed within a wider context, as was the case for the first one.

Box 9-6.

Relationships between the tool and object dimensions of the derivative in the project
(Artigue & al 1998, p.98)

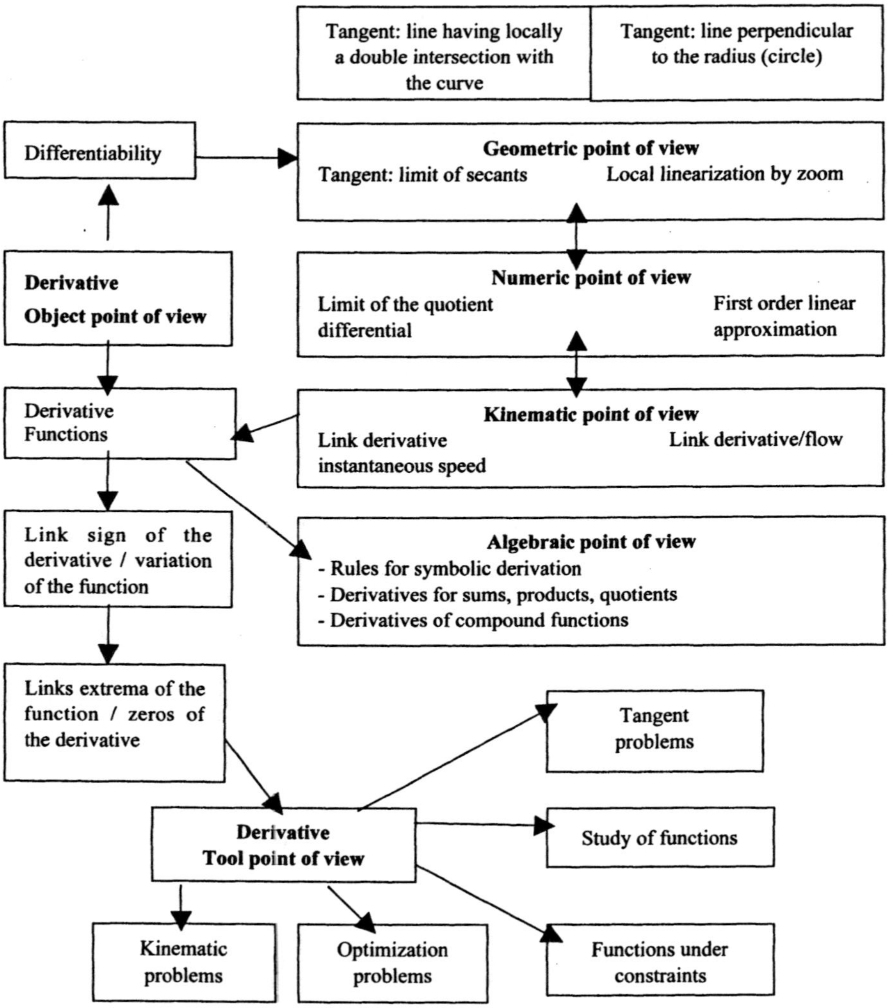

3.1.1 Exploiting graphic visualizations for problematizing the notion of tangent and establishing the local character of the notion of derivative

According to the point of view taken on the notion of tangent, the potential for visualization offered by the calculator can be exploited in different ways. One can use the geometrical software of the TI-92 in order to visualize the movement of a secant having a fixed point on the curve, when the second point of the curve which defines this secant tends towards the fixed point. One can also use the calculator to illustrate the fact that the graphic representation of a function at a point where it admits a derivative, tends to become a straight line when successive zooms-in on this point are performed. The first type of visualization leads to the tangent being perceived as the limit of secants and the derivative as the limit of differential quotients; the second type, to it being perceived in terms of linear approximation. But, as stressed, if visualization brings out some interesting perceptual phenomena, its does not provide their mathematization. In many innovations and experiments with graphic calculators, this question of mathematization is quickly passed over: functions and windows are chosen so that the line that appears on the screen has a particularly simple equation (with whole number coefficients) (Chapter 4, § 3), and the exact nature of this line is not discussed. This is treated as if it really were the tangent to the curve. The use of symbolic calculators, by the help they provide for the development of the computations underlying the mathematization, *modifies the economy* of the didactic system and *makes accessible* discussions on this mathematization which are not viable in the usual teaching environments.

3.1.2 Exploiting devices for data capture to allow a kinematic approach of the notion of derivative connected to students' movement

It is now possible to connect graphic calculators to devices for data capture (Chapter 4, § 2, see data logging device). In this way for example, it is possible to ask to a student to move, faster or slower, steadily or not, so that s/he gets back a discrete sampling of data. Given a time/distance curve, students can also be asked to imagine and carry out, if possible, a movement which corresponds to it. In the case of simple movements, the statistical module of the calculator can also be used to find, via the search for linear or quadratic regressions, equations fitting these data and to work on these equations. These devices allow a kinematic approach to the notion of derivative. They can serve to introduce this notion or, when this notion has already been introduced, for example by appealing to visualizations such as those described above, for connecting the notion of derivative to that of the instantaneous speed. This potential was not available on the first TI-92 used in this experiment and thus could not be exploited. Nevertheless, it seems important to mention it because it corresponds to cognitive approaches which are gaining increasing influence in the educational field: those related to the field of *embodied cognition*. Embodied cognition insists on the role played by our physical experience of movement in mathematical conceptualization (Lakoff & Nuñez 2000). For experiences drawing on these approaches with symbolic calculators, the reader can refer for example to (Oldknow & Taylor 2000).

3.1.3 Exploiting the calculator to approach differentiation in its symbolic dimension and develop an 'algebra of derivatives'

As was the case in the engineering design on numbers and algebraic expressions, differentiation provides the opportunity to approach symbolic computation as an *object*, not only as a *tool* (Box 9-1). Here, it is through the algorithmization of the computation of derivatives that this approach arises naturally. Two didactic strategies can be used for this purpose. In the first one, similarly to the first engineering design, the symbolic calculator is used as a provider of symbolic expressions and the students are asked to make sense of these expressions. In the second one, more ambitious, students are asked to program a module for the symbolic computation of derivatives.

In research carried out on formal calculus in secondary education, the second strategy seems more or less absent. This is easily understandable. One can find an outline of the type of mathematical and instrumental work necessary to underpin such a construction by looking at the module developed by Fortin for teacher training sessions organized by Texas Instruments[11]. This module exploits the already mentioned *Part* command for decomposing functional expressions into atomic elements, and then applies derivation rules to the generated tree[12]. Such a strategy is of problematic viability at high school level today, even in experimental circumstances, at least if one wants to develop it within the ordinary activities of the class. Beyond its evident cognitive cost, it appears too distant from the values that curricula emphasize today. However the situation could change if mathematics curricula were to be more sensitive to the connection between mathematics and computer science, as for instance the report on this theme by the CREM (Chapter 3, § 1) proposes, and as suggested in Chapter 1 (§ 2.3).

The first strategy can itself take different forms. In some cases, it is used to introduce the notion of derivative which thus appears first in its symbolic dimension. Here, the software works as a *black box* which transforms functions into other functions and what is at stake is the determination of the rules which govern these transformations. Linking this symbolic perspective on the derivative with others is then postponed. In most innovation and research projects, however, the notion of derivative is first established as a local phenomenon, using one of the strategies described above, and the black box device is used after this first phase, in order to set up the algebra of derivatives. An essential difference between these two strategies lies at the level of the *technological and theoretical* discourse involved in the corresponding praxeologies, according to the anthropological theory, evoked in Box 5-1. In the second case, even if the aim of the situation is the search for symbolic rules, reasons for the necessity of these rules, once discovered, can be sought on the basis of already built knowledge. In the first case, rationality can only be found, at this first stage, in the internal coherence of the formal game, and the development of a richer technological and theoretical discourse, from a mathematical point of view, is necessarily postponed. Doubtless, both strategies also differ at the level of prediction and checking. Analysis of the first engineering design showed the crucial role played by abilities of prediction and checking when the calculator is used as a producer of phenomena. It is evident that a first meeting with the world of derivatives, previous to the black box symbolic game, is likely to *change* the students' means of *prediction and checking* in that game, making this kind of situation more productive from a mathematical point of view.

3.1.4. Exploiting the symbolic potential of the calculator to work on more complex objects and to approach generalization

In this domain as in many others, the symbolic potential of calculators can serve to approach more complex situations or, at least, situations less calibrated and adapted so as to be compatible with the paper-and-pencil environment. But here we would like to insist more on another aspect: the potential offered by symbolic calculators for approaching issues of generalization. Analytic work, at high school level, deals with particular objects. It is carried out on specific objects and not on objects defined by general conditions, and the first access to generalization allowed by the use of parameters often seems incompatible with the students' algebraic competence. Nevertheless, overcoming the limitations of a perspective restricted to particular, isolated objects, is necessary to approach the connections, generalizations and unifications which are an essential component of mathematical learning. The symbolic assistance supplied by symbolic calculators should *here again change the ecology* of the system by modifying the cost of symbolic computations and of generalization through introduction of parameters, as already underlined in Chapters 5 and 7.

Thus, for this kind of engineering design as for the first one, symbolic calculators offer interesting didactic potential in principle. How was this potential exploited in the particular engineering design we analyze? With what effects? This is what we will now examine.

3.2 The engineering

We first describe the main choices of this piece of engineering, before focusing on some of its key situations.

3.2.1 The main choices

Having regard to what has preceded, this particular engineering can be described globally in the following way. The introduction to the notion of derivative exploits the first strategy listed, using the notion of tangent. The choice is made to use the geometrical module of the calculator and to visualize the tangent as limit of secants (Box 9-6), but the approximation point of view is also introduced very early. A progressive devolution of algebraic computations to the machine is organized and it is only in the third phase, when the notion of a derivative function has officially been introduced, and the cases of some *reference functions*[13] have been dealt with, that the symbolic dimension is introduced, by using the black box game outlined earlier. In the fourth phase, the relation between the sign of the derivative and the variation of a function is introduced by drawing on graphical experiments, and the corresponding knowledge is then used in solving simple problems of variation and optimization in the fifth phase. The sixth phase *comes back to the local point of view.* The hypothesis is made that this point of view, even though the initial one, tends to gradually fade from students' consciousness, replaced by the global point of view attached to derivative functions, which supplies them with a new and effective tool for solving the problems of variation and optimization which they meet in high school. This coming back is linked to the development of a *kinematic* perspective and the transition from the notion of average speed to that of instantaneous speed. Finally, the seventh phase of the process is that of reinvestment in more complex problems requiring connections between graphic and symbolic work, as well as generalization.

In what follows, we analyze four particular situations of this engineering, allowing us to discuss the exploitation of three different potentialities mentioned above: the situation of introduction of the derivative, the black box situation, and finally two situations that involve some generalization work.

3.2.2 The introduction of the derivative

The session aimed at introducing the derivative which we describe and analyze here lasts for 2 hours and is conceived in terms of three successive phases where the same problem of mathematizing a perceptual phenomenon is reworked with different points and functions. This aim of this device is to gradually increase the students' mathematical autonomy, and also to organize the transition from paper and pencil computation to instrumented computation, this transition being accompanied by a progressive increase in complexity of the computations concerned. We quickly describe the *a priori* scenario then give some information about the actual implementation, before coming to the *a posteriori* analysis.

The a priori scenario

Phase 1

The first phase focuses on the function $x \mapsto f(x) = x^2 - 2$, studied in the neighborhood of the point A (1, f(1)). The management is collective. The teacher manipulates the overhead-projected calculator. Using a prepared program, the teacher moves the variable point M on the parabola and thus the secant (AM). The coordinates of M and the slope of the secant are displayed in the upper-left corner of the screen (Figure 9-1).

The question posed to the students is the following one: *"What happens when M gets closer to A?"* Note that, when M is at A, the software sends back the value *undef* for the slope. In this situation one can expect students to identify the geometrical movement of the secant and to guess, in view of the trend of the slope, that the secant tends towards the line passing through A with slope 2, due to the attractive character of the whole-number value 2.

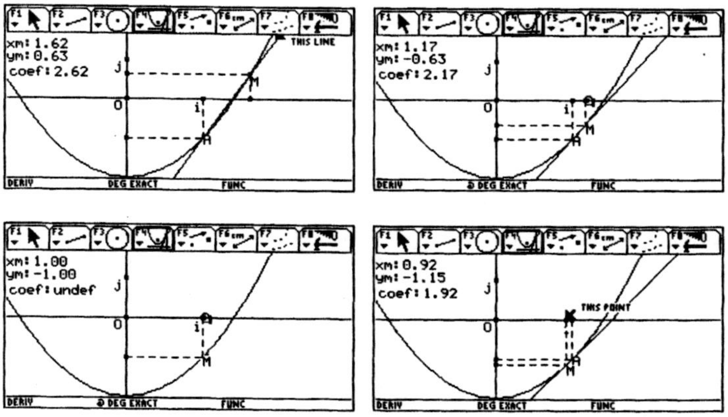

Figure 9-1. The movement of the secant in the geometry application

The scenario plans to then use the graphic application to question these interpretations. Entered in advance, in the graphing application *Y* =, are the equations of three different lines passing through A, with respective slopes: 1.9, 2, 2.1. To facilitate interpretation, the equations of these secants are not simplified but left in the form $y_i = y_A + m(x - x_A)$. In the graphical window, all the lines appear to fit equally to the parabola. This observation should allow the teacher to introduce a mathematical discussion on the legitimacy of the choice of the value 2 among the neighboring values (Figure 9-2).

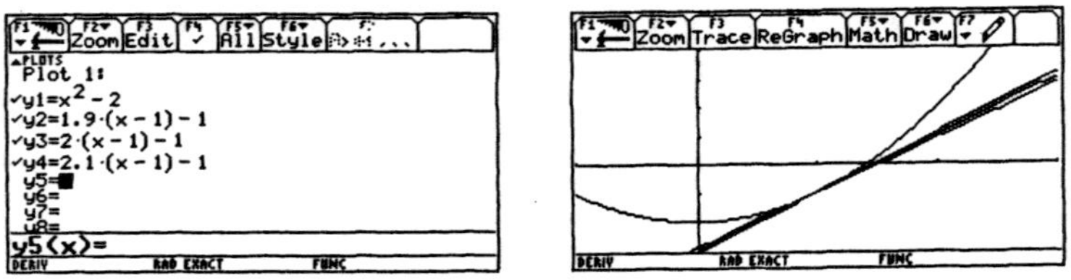

Figure 9-2. The three different lines

This graphic phenomenon persists under zooming-in around A which might be envisaged as separating the different lines, and this is intended to motivate the necessity of *going beyond perceptual criteria* to find the line best fitting the parabola. But, it seems unreasonable to expect grade 11 students to discover by themselves a technique allowing them to solve this problem, in the short time available here. That is why, in the scenario, once the problem has been articulated and understood, the teacher is expected to take more responsibility, and to suggest that the problem can perhaps be solved by studying the distance $\overline{P_hM_h}$ with M_h having x-coordinate $1 + h$ on the curve and P_h the points with the same x-coordinate on the three lines.

The scenario plans a collective management of the first computation, then a distribution of the other two computations between two halves of the class, to favor an individual appropriation of the collective work. This leads to the three expressions: $h^2, h^2 - \frac{h}{10}, h^2 + \frac{h}{10}$ all of which tend towards 0 as h tends towards 0, but it is expected that the students, thanks to the previous work they did on limits and orders of magnitude, will be able to distinguish these three expressions according to the respective speed with which they tend towards 0. Thus the value 2 appears particular: for it $\overline{P_hM_h}$ is has no term of order h, or equivalently: $\lim_{h \to 0} \frac{\overline{P_hM_h}}{h} = 0$. It then remains to show that it is the only value to have this property. The corresponding computation necessarily involves a parameter but it reproduces a type of computation already done three times with particular values. It is planned to carry it out collectively, piloted by the students, and by hand with paper and pencil, as the preceding ones:

$$\overline{P_hM_h} = f(1 + h) - [f(1) + mh]$$
$$\overline{P_hM_h} = h^2 + h(2 - m)$$

This last equality shows that $\overline{P_hM_h}$ has no term of order h if and only if $m = 2$.

At this point, the scenario incorporates a first summing up of the work piloted by the teacher and the introduction of the word *tangent* if it has not already emerged spontaneously (these students have previously solved a problem on the tangents to a parabola, by adopting an algebraic point of view: looking for double intersections by analogy with the tangents to the circle). This connection with the former point of view on the tangent concludes the work on this first example chosen so that the computation does not present any difficulty (since it leads to a well known identity).

Phase 2

In this phase, the same method has to be applied, this time to the function $f: x \mapsto f(x) = \dfrac{x^3 - 6x}{4}$ at the point with x-coordinate 2.

For this adaptation to a new function, the plan is for the students to pilot the solving process collectively, the calculator being used for computations, with the various steps involved as well as the details of the computations and the results obtained being noted on the blackboard. This example has also been chosen to make students meet a case where the tangent cuts the curve (but not at the tangential point).

The question posed is the following one: *"Is what we found with the parabola still valid here and do the same methods of computation work?"*

This is, of course, the case, as shown by Figure 9-3.

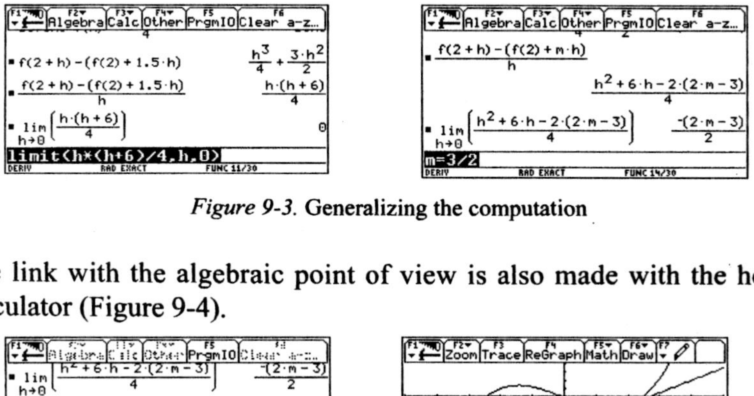

Figure 9-3. Generalizing the computation

The link with the algebraic point of view is also made with the help of the calculator (Figure 9-4).

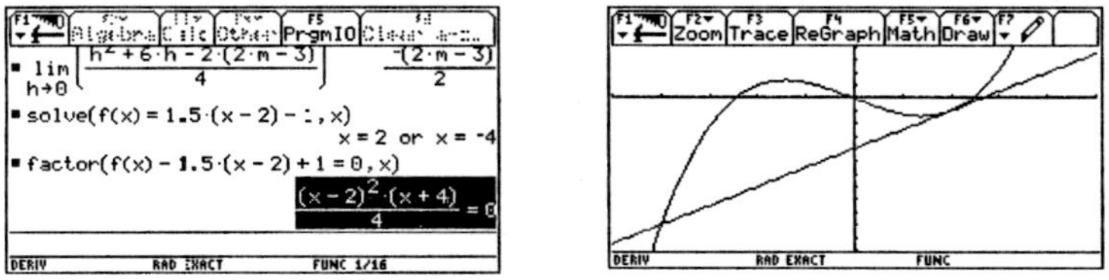

Figure 9-4. Link with the algebraic perspective

Note that the search for intersection points with the *Solve* command does not make it possible to distinguish between the two roots: 2 and –4, but that this can be achieved through the use of the *Factor* command.

The plan is to end this second example with a summing up where the fact will be stressed that tangents to a cubic curve can cut it, contrary to what had been noticed for the parabola.

Phase 3

Working more autonomously, in pairs and with the calculator, in this third phase the students themselves have to handle the case $x = 0$ for the same function. The hypothesis is made that the whole method will then be accessible to them. This new case will confront them with a situation of inflection, leading to discussion of the doubtless strong conviction that a tangent cannot cut the curve at its contact point.

As a conclusion to these three phases, it is planned that the teacher introduce the definition of the derivative of a function f at the point with x–coordinate a, and stress the equivalence between the two following formulations:

For some m,

$$\lim_{h \to 0} \frac{f(a+h)-(f(a)+mh)}{h} = 0 \text{ and}$$

$$\lim_{h \to 0} \frac{f(a+h)-f(a)}{h} = m$$

which makes it possible, by directly seeking the limit of the differential quotient, to simplify the work of guessing and proving carried out on the first three examples.

The implementation

This scenario of introduction was experimented in the following way: a collective session of one hour (phase 1) followed by a one hour session of work in half-groups (phases 2 and 3). What emerged from this experiment?

The beginning of the session took place according to the predictions made and no student introduced the word *tangent*. The students seemed surprised by the closeness of the three lines and even more by the fact that the zooms they proposed did not make it possible to clearly discriminate the line with slope 2. As anticipated, they had no idea *a priori* about possible techniques for comparing the respective nearness of the three lines and the curve. Thus, the teacher suggested studying the value of $\overline{P_hM_h}$ and introduced the notations, as expected. The expression of the algebraic value of $\overline{P_hM_h}$ in the case of the line of slope 2 was produced collectively, and set out by a student, after a very brief period of individual work. The computations and various substitutions were carefully detailed. The teacher asked: *"What does this expression become as h tends towards 0?"*. Various answers were produced: *"It becomes nothing"*, *"It is invalid"*, *"It tends towards 0"*.

The teacher summarized and asked for a reformulation in the usual language of limits, underlining that the function of h under consideration is a reference function (note 13). The class was then divided into two groups to deal with the two other cases. This computation was carefully prepared: the teacher asked where changes in the algebraic expression were going to occur, and the work was quickly executed by the great majority of the students. They seemed disappointed to notice that all the expressions had the same limit 0, which was nevertheless predictable, thinking doubtless that the suggestion made by the teacher should lead directly to the result. The three results were written on the blackboard in a table, the teacher ensuring that the computations of limits were carefully justified, which apparently raised no problems for the student being questioned.

Having found these limits, the teacher asked how to continue. This time she obtained an answer at once: *"We could see which one goes the fastest"*. The teacher immediately reformulated this suggestion as the absence of a term of order h, appealing to the memory of the class. A column was added to the initial table to record the limits of the quotients, $\overline{P_h M_h}/h$, and the limits of these were found and justified without any difficulty. The teacher then asked the students to write down this conclusion in their exercise books, before relaunching the problem, as expected, by asking the question: *"Is the line of slope 2 the only one with this property?"*.

A student immediately suggested redoing the computation in terms of x, (x being, for him, the slope). The teacher suggested using the letter m to avoid confusions and let the students do the computation. Apparently, for most of the students, the structure of the expression was well understood and the adaptations quickly made. However one student asked if $f(1+h)$ was equal to h^2+2h-1 and the teacher went over the substitution in detail with the class. Thus they arrived at the expression $h+2-m$ and quickly concluded that 2 was indeed the only possible value. The teacher recapitulated the different steps in the reasoning and noted the result obtained, before concluding: *"Thus this line plays a very special role with regard to the parabola at the point with x-coordinate 1"*.

A student asked: *"Is this the derivative?"*. The teacher asked him to make his suggestion clearer and the word *tangent* was introduced by several students. The teacher immediately took advantage of this to ask on what occasions they had already heard about tangents. Some spoke about the tangent to the circle as being perpendicular to the radius, of the tangent in trigonometry. Appealing to the memory of the class, the teacher asked them if they did not remember work done earlier that year. Students then recalled the homework mentioned above. The teacher brought out the link with the present situation by returning to the geometrical visualization and made them recall the strategy used in the problem: writing a system of equations to determine the intersection of a line and the parabola, solving the system and looking for double roots to find the tangent. The students then carried out the computation in the case of the parabola and quickly found that the double root was 2.

The teacher then recapitulated the whole activity, announced that this slope of the tangent was going to play a very special role and that, in the next session, they would have to do similar work with other functions.

During this session, in the first half-group, what happened at the beginning conformed to the scenario. The limit value 1 obtained on the screen perturbed some students who wondered whether the line of slope 1 was really the closest to the curve in the neighborhood of 1. The teacher showed the line of slope 1 by moving the point M and the conjecture was quickly refuted. The adaptation to this new situation was apparently made without difficulties, the students piloting the collective computation while carrying it out at the same time on their machines. After the characterization of the value 1.5, the teacher came back to the graphing application *Y* =, and asked them to study more generally the respective positions of the tangent and the curve. As expected, the fact that the tangent cuts the curve (but outside the window used initially) perturbed many students who began to doubt the reliability of the tangent. The teacher did not let them persist in this difficulty and asked those who had plunged into a graphic investigation of the situation to return to the algebraic computation.

Roots were found with the *Zeros* command, or the *Solve* command (Figure 9-5).

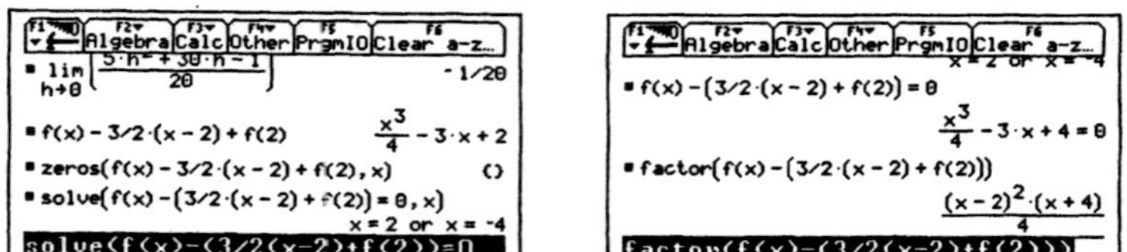

Figure 9-5. Finding the roots with the symbolic calculator

Several students were amazed to find there were only two roots for a polynomial of degree 3, and the conjecture was soon made that there was both a double root and a single root, the double root being 2. The teacher asked how this conjecture could be verified and a student immediately suggested factorizing. The expression obtained satisfied everybody and the problem was apparently closed.

The teacher commented on this situation, pointing out the difference with the parabola and insisting on the fact that the property of tangency is a local property: a line is tangent to a curve at a point.

As expected, she then asked them to study individually the case x = 0 for the same function, without trying this time to make them collectively form a conjecture about the slope of the tangent. Most of the students undertook a graphical investigation, reducing the standard window so as to better see the curve in the neighborhood of the origin, but the time remaining was less than 10 minutes and the teacher very quickly resumed a collective management.

One student proposed at once that the tangent had slope 1.5. He had found this value by using the *Derivative* command of the menu *F5* which had not been officially introduced. The teacher showed them that there was also a *Tangent* command which gives a value different from 1.5, and took this opportunity to stress the fact that the computations performed in the graphic application are always approximate even if the exact mode has been selected (Figure 9-6).

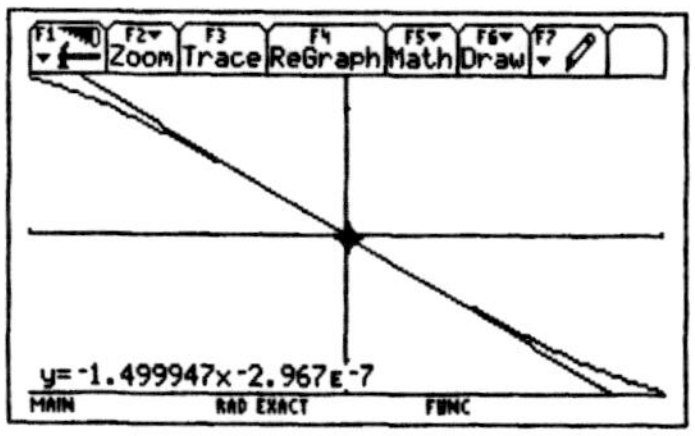

Figure 9-6. Using the *Tangent* command

She then asked the students how to decide whether 1.5 is the right value or not. Students suggested calculating the limits with respect to 1.5. The teacher accelerated the process by suggesting a direct computation with the parameter m. The session ended with this computation, and with the teacher saying that in the following session she would define the value of the derivative by drawing on what had been done during this session.

In the second half-group, the teacher met a small technical problem during the investigation: she could not separate points M and A, having released the mouse while both points were very near to one another. Not succeeding in solving this technical problem quickly, she jumped to the application *Y* =. The dynamics of the session were then approximately the same as for the first group; for the first case, however a bit faster.

In the second case, after graphing the curve, a student declared that this could not have a tangent at 0 (showing the inflection with a hand movement). Others did not agree with him. The teacher then suggested using the *Tangent* command in the menu *F5*. The students noticed that the calculator gives the equation of a tangent, and that this tangent cuts the curve at 0.

Some found this strange: " *It looks like a tangent, but it crosses*". Some of them were really perturbed. The teacher then asked them to count intersection points. Most answers were that there was only one intersection point but a student suggested that it could count as 3. The teacher asked for a verification and the same student suggested using the same method as previously with a point P and a point M. This reorientated the work towards the classic strategy and they arrived at the end of session at the conclusion that $\overline{P_h M_h}$ has no term of order h if and only if $m = -3/2$.

The teacher concluded that this line was going to be the tangent to the curve at the origin, even though cutting it, and that the number m was going to be the value of the derivative of the function at 0. She also introduced the *Derivative* command of the menu *F5* which had been used spontaneously by a student in the other group. She asked them to look for the number of intersection points between the tangent and the curve in preparation for the next session.

Analysis

This session differs in many respects from the sessions presented from the first engineering project. On the one hand, prior to it taking place, the students had already developed some instrumentalization and instrumentation of the TI-92; on the other hand their mathematical knowledge was different from that of the grade 10 students. Their capacities with regard to the *identification of forms* needed to interpret symbolic computations and to substitute other numerical or even literal values seemed effective enough to allow them to access the individual and collective work central to this session. The *differentiation between different orders of magnitude*, which had been introduced in preliminary work on limits, was easily brought back into use in this situation where the comparisons remained technically easy. *The problem of mathematization* of an interesting perceptual phenomenon seemed to make sense for these students who could have been considered *a priori* as less interested than others in mathematical abstractions since most of them planned to enter vocational programs. But what we want to stress is that what we observed, three months after the introduction of the TI-92 to this class, were the effects of an instrumentation process which had been organized and coherently managed by the teacher, and of a *culture of instrumented mathematical work that had been progressively built up.*

The careful management of the variables affecting the complexity of the computations concerned doubtless played a decisive role too, as did the fact that, during the session, on different occasions, students had to reproduce, modulo some adaptations, similar computations, and also the progressive interaction of work instrumented through pencil-and-paper and through the machine. Such *sensitivity to the complexity* of the computations when working with symbolic calculators or software is not very frequent. Misled by our personal experience, we are tempted to forget that, in order to benefit from the machine, the student *must give some sense* to the computations s/he is asked for, which is not so easy as the symbolic complexity of the computations increases. The report (Artigue & al 1998, p.19) shows the deceptive results obtained during the first year of experimentation as we were not sensitive enough to these complexities. A reflexive analysis on this first experiment allowed us to become aware of their importance and consequently to adapt the design.

Another essential difference to the first project is related to the management of the session. This situation cannot be qualified as an *adidactic* one, in the particular sense given to this term in the theory of didactic situations (Box 9-7). We cannot say that students are faced with a problem and solve it progressively thanks to the interactions they have with the *medium* (Box 9-7). A problem can certainly be identified: to make sense mathematically of a certain type of numeric-graphic phenomena, but this problem breaks down into a succession of sub-problems, in a scenario *tightly piloted* by the teacher. Choices are thought to give, in every phase, the *maximum autonomy* to the students but the students' and teacher's lines of action are strongly intertwined, precisely to allow an autonomy that the first experiment had shown to be something difficult to achieve. Here the role of the teacher is decisive at each point and, to make this visible, we adopted a different style, more narrative, to present the scenario and report on the actual implementation. Of course this raises the question of the students' personal commitment and of the learning opportunities that each of them was able to seize or not. The first session, essentially collective, does not make it possible to answer this question. Rather, what was tested there was the *viability* of a collective activity; as regards individuals, we only have the information provided by the collective exchanges about those students involved in them. However, the data collected during the two half-group sessions where four students at different academic levels were specifically observed, tend to confirm the impression that the introduction provided by the collective session was viable.

The scenario which was built presents however evident weaknesses. Firstly, it is too long for only two sessions, if the expectations of the *a priori* analysis as regard the division of mathematical responsibility between teacher and students have to be respected. In both groups, the last part was not reached and the phase 3 was covered too quickly and collectively. Furthermore, as underlined in the report, the scenario of the third phase does not ensure that students will find in their interactions with the medium the means allowing them to carry out the mathematical work expected. Indeed, the students do not have a personal version of the program used in the first two phases in order to provoke conjectures. This was only installed on the teacher's calculator and used collectively with the view-screen. Thus what is asked of them, is not the mere reactivation of a previously used strategy, as they might expect. In this context, two strategies seem *a priori* possible:

- by making successive zooms or by choosing a suitable window, to produce a nearly linear graph and to exploit it in order to conjecture the slope of the tangent as a simple decimal number;
- to directly undertake a computation involving the parameter m.

Box 9-7.

Adidactic situations and devolution
(Brousseau 1997)

In the theory of didactic situations, elaborated by Brousseau, the notion of *adidactic situation* plays a fundamental role. In such situations, the teacher provokes the adaptations expected of students by a clever choice of the problems s/he sets up. Between the moment when the student accepts the problem as his or her own and the moment when s/he produces an answer, the teacher refuses to introduce the target knowledge: the student works at solving the problem, alone or with other students. Knowledge is produced through interactions with the *medium* which is defined as the system in opposition to the student, and which can include both material artifacts and symbolic tools. Such a situation is called *adidactic* because its teaching aims remain invisible in an important sense; nevertheless it is organized with an undeclared didactic intention. The teacher wants to devolve to the student an adidactic situation which will induce the most effective interaction possible. The strong didactic contract is temporarily relaxed permitting the student to be modeled as a "mathematical subject". Setting up adidactic situations is not easy and Brousseau introduced the word *devolution* to label the process by which the teacher devolves mathematical responsibility to the students and tries to maintain it. One of the basic principles of the theory of didactic situations is the following: real mathematical learning requires adidactic situations or at least adidactic phases in didactic situations.

The designers of this piece of engineering thought that the first strategy, even if requiring some initiative, was reasonably accessible and that the second one might appear but would certainly remain marginal since it required a real change in perspectives. But they planned to exploit the difficulty met by the students in this phase to introduce the definition of the derivative and make the students sensitive to the value of the second formulation (see above). As shown by this experiment, these predictions did not work: in the first group, because students made use of specific commands offered by the menu *F5*, but not yet introduced; in the second group because the situation provoked a debate on the existence of the tangent, in the case of inflection. We cannot say what would have occurred if the students had had more time to solve this problem, and if the teacher, under the pressure of time, had not stopped individual work so quickly. The question remains open.

3.2.3 The formal computation of derivatives

From the start of the experimentation, the teaching of rules for calculating derivatives exploited the symbolic potential of the calculator. We asked students to observe how the TI-92 answers when commands asking for the derivative of sums, products, quotients of general functions (for example: d(f(x) + g(x), x), d(f(x)/g(x), x), or composite functions are entered (a similar task about limits is presented in Chapter 3). They had to transpose what appeared on the right part of the screen in terms of $\frac{d}{dx}$ into the standard notations they used for the derivative and to conjecture rules which were then discussed collectively. The results were not convincing. As in the engineering with grade 10 students, the work of observation and transcription was of limited productivity, and insufficient to *motivate the search* for reasons for the conjectured rules. Furthermore, it appeared that when the calculator simply returned the original expression used to calculate, for instance with composite functions, the students had difficulties of interpretation and tended to infer erroneous rules, such as: $(f(kx))' = f'(kx)$.

The engineering was thus modified for the second year. More precisely, we planned:

- to ask for predictions before using the calculator;
- to work first on some particular functions, in order to establish a basic repertoire of associations between functions and derivatives, which would help to *motivate the search for more general rules.* Looking at the functioning of the TI-92 with general expressions was postponed to a second phase, and linked to the development of a more theoretical discourse aiming at finding mathematical reasons for the rules observed.

In this situation students have limited means of prediction. However, one can reasonably expect that they will extend the conservation of operations in relation to the algebra of limits into an algebra of derivatives, leading them both to correct (for sums and multiples of functions) and incorrect (for products, inverses, quotients of functions) conjectures. The confrontation between students' predictions and the results given by the calculator should thus be particularly interesting.

Scenario

The session is planned in two phases, separated by a summing up, for a total duration of 2 hours. In the first phase, the aim is to establish rules for the computation of derivatives, with the help of the TI-92, as explained above. The derivatives for reference functions already learnt are first recalled collectively. The student worksheet is reproduced in Box 9-8.

Box 9-8.

Student worksheet, first phase

(Artigue & al 1998, p.165)

Computation of derivative functions

Your TI-92 calculates derivative functions quite well:

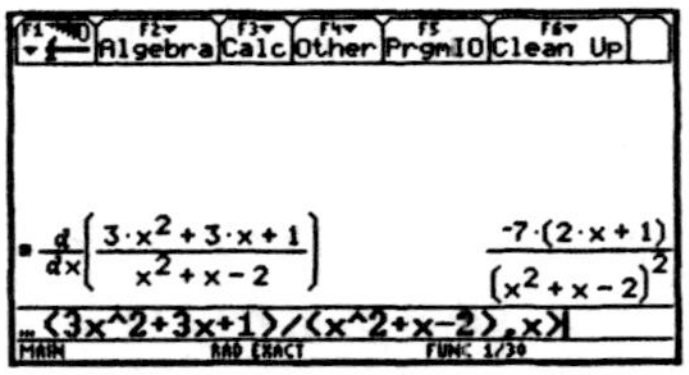

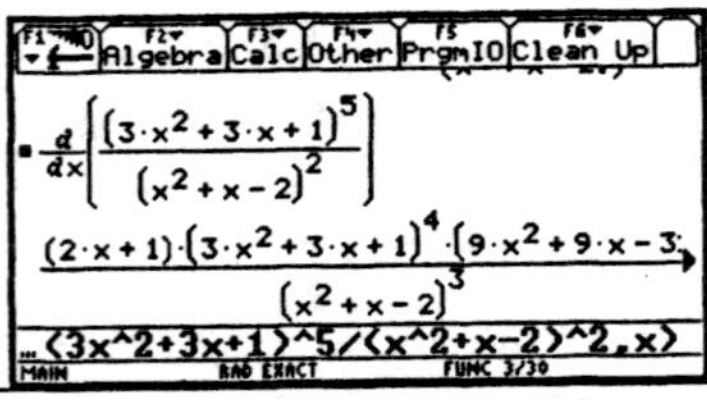

You can use the menu *F3* then 1: *Differentiate*, or [*d*], or [*CATALOG]*	Don't use the normal style of d on the keyboard and don't forget the variable for differentiation

Recall the results found on Monday and check these with the machine:

Function defined by	Derivative function
$x, x^2, x^3, \sin x, \frac{1}{x}, \sqrt{x}, b.$	

How can you be nearly as fast using paper-and-pencil?
As we did with the limits, we will look for rules:
Fill the following table by using your imagination:

	Function defined by	Derivative function
1.	x^4	
2.	$x\sin x$	
3.	Other functions: 2x+1 ; sin(2x) ; 3sin(x) ; 4sin(3x) ; sin(x+1) ; cos(x) ; cos(2x) ; f(x)+g(x) ; f(x)g(x) ; $\frac{1}{f(x)}$; $\frac{f(x)}{g(x)}$.	

Now using your TI-92, can you confirm the above rules?

	Function defined by	Derivative function
1.	x^4	
2.	$x\sin x$	
3.	Other functions: 2x + 1 ; sin(2x) ; 3sin(x) ; 4sin(3x) ; sin(x+1) ; cos(x) ; cos(2x) ; f(x) + g(x) ; f(x)g(x) ; $\frac{1}{f(x)}$; $\frac{f(x)}{g(x)}$.	

In the summing up, it is planned to list the rules discovered by the students, to discuss these, to search for justifications by relying on limits, and to discuss the functioning of the TI-92 and what can be expected from it, especially through analysis of the answers given to the derivative of general expressions (this analysis shows for example that the quotient f/g is treated as the product of f with 1/g).

Finally, in the second phase, students are asked to practice on various examples, with the computations being performed by hand, and the calculator serving to check these hand computations. The worksheet distributed to students (reproduced in Box 9-9) asks them to indicate the rules that they are using, and the interval on which the computation is valid, determination of this interval being, even in case of instrumented work, the responsibility of the student.

Box 9-9.

Student worksheet, second phase

(Artigue & al 1998, p.167)

Summing up of results		
	Function defined by	Derivative function with the TI-92
1.	f(x) + g(x)	
2.	f(x) g(x)	
3.	$\frac{f(x)}{g(x)}$	
4.	$\frac{1}{f(x)}$	
5.	kf(x) with k a constant	
6.	f(kx) with k a constant	
7	x^n with n an integer	
8.	f(x + k)	
9.	cos(x)	

We will prove these rules in the next session; they result from computation rules for limits.

Using differentiation rules

By using the previous rules, calculate the derivatives of the following functions, and state for what values of x_0 this computation is possible. Check with the machine.

Function defined by	Derivative function	for x_0 in	Rule used
$x \mapsto x^7$			
$x \mapsto (3x+2)(2x+3)$			
$x \mapsto 3x^4 + 5x^3 - \frac{2}{3}x^2 + 5$			
$x \mapsto \frac{x+1}{x-3}$			
$x \mapsto 1 + \frac{1}{x} + \sqrt{x}$			

$x \mapsto \sqrt{x+1}$			
$x \mapsto \dfrac{x^2+3x+1}{x^2+1}$			
$x \mapsto (x+1)^2(x+3)$			
$x \mapsto \sin x.\cos x$			
$x \mapsto \sqrt{2}(x^3-x+7)$			

As can be observed, trigonometric functions are over-represented in the corpus of particular functions: 7 out of 9 (reproduced in Box 9-8, 3). We hypothesize as follows: these functions are well adapted to the work on derivatives of composite functions which is planned later on in the session. In fact, four cases among these first nine are already composite functions. On the contrary, rational expressions which correspond to the cases which students meet most frequently at this level are under-represented (2 cases only) and there is no quotient. Thus the set of functions seems unbalanced with regard to what could be expected *a priori*. With the exercises of the second phase, we meet the opposite situation: only one trigonometric function is proposed and rational functions strongly predominate (7 cases among 10). The technical work focuses thus on the categories of objects whose mastery is expected.

Implementation

The first worksheet (Box 9-8) concerning the *reference functions* (note 13) was collectively filled in after a short introduction by the teacher. When the students came to check their answers with the calculator, some of them obtained a different result for the *sine*. The problem resulted from the fact that their calculator was in degree mode and/or in approximate mode. The teacher took advantage of the occasion to remind them that the derivative of the function *sine* is the function *cosine* only if x is expressed in radians. However, at this time, he could not yet explain the multiplicative coefficient which appeared by appealing to the derivative of composite functions.

The students then proceeded to the work of *prediction,* the teacher asking them to switch off their calculators and to try to justify their predictions. Box 9-10 below summarizes what they proposed after the ten minutes devoted to this activity[14].

Box 9-10.

Summary of the students' predictions

Functions	Comments
$x \mapsto x^4$	Correct derivative for most students, found by using the previous examples (x^2 and x^3). One student writes x^4 under the equivalent form $(x^2)^2$ and calculates the derivative by following a rule $(fg)' = f'g'$.
$x \mapsto x\sin(x)$	Two students give the correct derivative *. The others return the product of the derivatives or the derivative only of the sine.
$x \mapsto 2x+1$	Fifteen students give the correct derivative. A student adds that the slope of 2x + 1 is 2, another one that the derivative of 1 is 0. Six other students add 1 to the derivative of 2x.
$x \mapsto \sin(2x)$	Four students give a correct answer*. Two others answer according to: $kf'(x)$, and the majority (11) according to: $f'(kx)$.
$x \mapsto 3\sin(x)$	Twenty students give a correct answer, one returns cosx and one 0; he followed a rule $(fg)' = f'g'$.
$x \mapsto 4\sin(3x)$	Five students give a correct answer (the same ones as for sin (2x) plus one more). Other students follow a rule $k f'(x)$ or $f'(kx)$) as above. A student (the same one as above) obtains 0.
$x \mapsto \sin(x+1)$	The correct answer is predicted by 10 students. Four others return the derivative of the 1 in the bracket as 0.
$x \mapsto \cos(x)$	Three students know the derivative of the cosine. The majority think that sine and cosine are exchanged by derivation.
$x \mapsto \cos(2x)$	Four students predict correctly the derivative of the cosine and the derivative of f (kx)*. Two students predict only the derivative of f (kx).
$x \mapsto f(x)+g(x)$	The rule is correctly predicted by the twelve students who answer.
$x \mapsto f(x)g(x)$	Only one student gives the right rule for the product. Ten other students assume the conservation of the product.
$x \mapsto \dfrac{1}{f(x)}$	Ten students are influenced by the derivative of 1/x and thus omit $f'(x)$. Three others return $1/f'(x)$. One takes the derivative of the numerator and obtains 0.
$x \mapsto \dfrac{f(x)}{g(x)}$	Few students tackle this last question. Two answers assume the conservation of the quotient and three are influenced by the derivative of 1/x, proposing $-f(x)/g(x)^2$.

In the summing up which followed, the teacher wrote on the blackboard the results given by the students without saying a word. He asked them then to redo the same work with the machine, so as to fill in Box 9-9. He pointed out that the machine may give false results and that they had to argue and validate the correct answer. Obtaining the results with the TI-92 did not raise any major problems. We noted only some problems of syntax: implicit product for x.sin x, previous assignment of the functions f, g in the general case, forgetting derivative signs in the transcription to paper.

The screens obtained are reproduced in Figure 9-7.

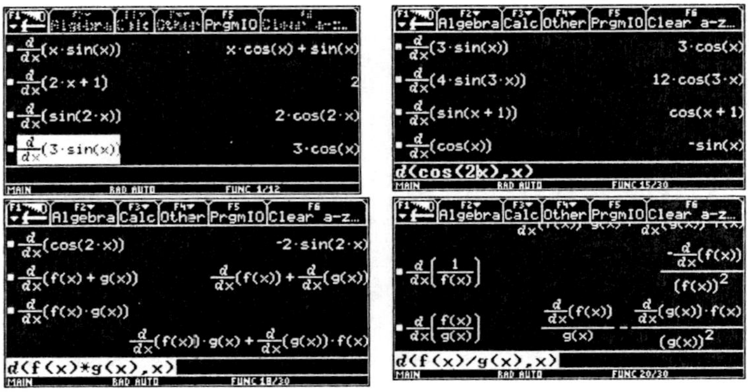

Figure 9-7. Derivatives on the TI-92

A little more than half an hour after the beginning of the session, the comparison between the conjectures and the results supplied by the calculator began, conducted collectively. It generated lively discussion between students and with the teacher, but the students often had difficulty in explaining their conjectures and in associating precise rules with these, which limited the scope of the discussions about the first examples (the particular cases). The transition to general expressions made it possible to return, as expected, to the specific cases by wondering about the consistency of the answers given to the particular cases with the general rules. It also made it possible to point out seductive errors. In the case of the inverse, the link was made with the derivative of $1/x^2$ and, in the case of the quotient, a collective computation was done: reducing both quotients to the same denominator in order to get the usual formula. The summing up was satisfactory, even if it remained more at a descriptive level than expected: students stated the rules and the consistencies but they did not go beyond this level; nevertheless when the summing up ended they seemed a bit tired.

Approximately one hour after the beginning of the session, the results were recorded in the third table and, with this table, new cases appeared: those relating to compound functions especially. Contrary to what might have been expected, the students had difficulty in conjecturing the derivative of $x \mapsto k.f(x)$ from the examples already met (did this result from tiredness?); so the teacher moved to the TI-92. But they also met difficulties in interpreting the productions of the machine correctly, linking these to the particular cases already treated. The teacher intervened firmly; he asked for 3 to be substituted in place of k (form |k = 3) so as to make them aware of the fact that the calculator does not give any answer when f is not specified.

For f(x + k), having reminded students of the particular case sin(x + 1), he asked for a graphic interpretation. In that case, indeed, a graphic interpretation is more accessible than in the case of products because only translations are involved, but this attempt, too abruptly introduced, did not generate any real recognition. The teacher closed the first phase at this point and asked the students to work individually for the second phase. He also announced that the proofs of rules would be done in the next session.

This second phase began after about one and a quarter hours of intensive work and the teacher insisted on the third column of the table being filled, that dealing with domains of validity. He would have liked his students not to reduce derivatives to their symbolic aspect. During the individual work he again insisted on this point several times, but unsuccessfully. In fact, the students had grasped the symbolic aspect of the work developed during this session and were not willing to accept another agenda. Essentially, their errors concerned the management of constants (5 remained in the derivative of expression 3, 1 disappeared in the derivative of expression 6, for example), the derivatives of products (for some students, the derivative of the first expression was 6, of the last one, 0) and quotients. There were also many algebraic errors in the expansions, which is not surprising if one considers the symbolic complexity of some of the expressions proposed. Finally, the students had difficulty in coordinating the results they obtained by hand for sinx.cosx, $\cos^2 x - \sin^2 x$, and as supplied by the TI-92 , $2\cos^2 x - 1$.

A posteriori analysis

This session, as noted above, aims at exploiting the productions of the machine to motivate and develop mathematical work on the rules of differentiation. This makes it closer to the sessions from the first project, as regards the functionality given to the calculator. It is also closer in terms of management: students work with worksheets defining a corpus of functions which will be used as a base for reflection and elaboration of rules. As was the case with grade 10 students, the limits of such an organization, when observations cannot be linked to a structured system of predictions, were quickly perceived as a result of the first experiment. The second experiment, on the contrary, tends to confirm the potential for mathematical work of the confrontation between predictions and results supplied by the machine, in a case such as this one, where the rules on which the prediction is more or less explicitly based, are partially erroneous. The predictions made by the students, summarized in Box 9-10, and the liveliness of the collective discussions clearly evidence it.

But observations also show us the *multiplicity of the variables* which intervene in the efficiency of such a session. First of all, the work of prediction can only be productive if machines are switched off. Here, this is the case when the activity is launched and the students, being accustomed to respecting this type of didactic contract in the classroom, do not try to free themselves from it; nevertheless as some students begin to fill the worksheet during the first phase of the session, there is some unexpected perturbation to the work of prediction. Another point is that the phase of prediction must have a real *status*. If it seems just like an introduction, the confrontation will lose much of its efficiency; the authority of the result supplied by the machine will sweep away all the previous work, whatever its coherence. In this situation, observations show that the students' predictions are consistent, at least locally. Trying to clarify some of these consistencies, and to pose questions about them before using the calculator gives the prediction work quite another status. Here, what actually unfolded does not seem to take such a form: *the time devoted to* prediction *is short*, the results are recorded on the blackboard without comment; there is no discussion either by groups or the class organized as a whole before passing on to the work with the machine. Even if the discussions are then lively, their productivity is limited by the fact that students no longer recall why they predicted such and such a thing and, *a posteriori*, they are certainly not motivated to look for the reasoning behind answers invalidated by the machine. The third variable is the variety of cases which one wishes to treat. Here, managing in the same session *the destabilization* of spontaneous rules concerning the differentiation of products and quotients (a case more difficult than it appears at first sight), the differentiation of constants and the differentiation of compound functions (even when limited to composition with linear functions), *this is too much for a single session*. Serious work on each of the objectives becomes difficult and one can easily slip into an approach which remains superficial. This limits in particular the investment that it is possible to make in the technologico-theoretical work outlined in (Box 5-1). Students notice more than they explain, they observe more than they argue about, justifications are postponed to another session, in spite of the initial intentions. To this must be added the delicate problems of interpreting the expressions provided by the machine in the case of composite functions, certainly premature at this point.

In the case of composite functions, we also see the teacher encountering serious difficulties on discovering the *incapacity* of the students to use the trigonometric examples they have previously met, trying to introduce a new and interesting way of justification, by coming back to the geometrical meaning of differentiation. But the students were unable to make sense of this proposal breaking with the symbolic line of the session, introduced at the level of a collective discussion, and without any preparation.

Taking this analysis into account, the engineering was modified once more. Two different 2 hour sessions were organized: the first one centered on algebraic operations on derivatives: sum, product by a scalar, product, inverse and quotient; the second centered on composite functions. In the first session, prediction of particular cases and conjectures about the four operations, as quoted above, formed the theme for work in small groups followed by a collective summing up. This summing up aimed to clarify general conjectures as well as their consequences in particular cases, and to reject in a reasoned way some erroneous predictions, to formulate some questions. The calculator was only used after this substantial phase. The search for reasons for the rules of differentiation of f + g and f.g was, in the summing up, *piloted* by the teacher relying on the *approximation* point of view. Then the other rules were formally derived from these by the students.

The second session, focusing on composite functions, was planned for later in the academic year. Trigonometric functions played a privileged role in it, as was the case in the initial engineering, but also the square root function which makes it possible to tackle questions related to the *domain of validity* of symbolic computations (Chapter 2). The work was not only conducted at the symbolic level. The graphic setting was also brought into play. Graphic representations of specific functions of the form $x \mapsto f(x+a)$ and $x \mapsto f(kx)$ for several values of a and k were used to establish the corresponding rules of differentiation. Proofs linked graphic and symbolic elements but remained guided by the teacher, in particular for the product.

This new design was not systematically observed in the way that the preceding one was, but, according to the teachers involved in the initial project who are still using this approach, it now works in a completely satisfactory way.

3.2.4 Accessing generalization

In Chapter 5 (§ 6), Lagrange discusses the potentialities of computer tools in allowing students access to generalization. He concludes that CAS provide students with new techniques which, when inter-related to old techniques, can open up new understandings and support generalizations which lie beyond the range of student possibilities at this level in paper-and-pencil work. In this section we consider this issue again, focusing on access to generalization through functional activities involving parameters, and on conditions for the viability of such activities at high school level. For that purpose, we use two situations from this piece of engineering. The first one -- the tank situation -- corresponds to one of the first optimization problems met by the students; the second -- the pipe situation -- is a much more complex situation involving functions defined by constraints. Posed initially to the students over two sessions at the end of the year, it was then integrated, under different variants, within a project involving Internet exchanges between classes, piloted by the inter-IREM ICT commission (Box 1-3). We briefly present these two situations and what we learnt from the experiments. The reader can consult the report (Artigue & al 1998) and the web site of the Academy of Besançon (http://www.ac-besancon.fr) for more details.

Box 9-11.

The tank problem
(Artigue & al 1998, p.54)

A mason has to make in concrete a tank in the form of a parallelepiped. The base is a square, the concrete has a thickness of 20 cm, and the interior volume of the tank is 4 m^3.
x (in m) denotes the internal measure of the side of the square base and h (in m) the internal height of the tank.
Determine x and h so that the volume of concrete be minimal.

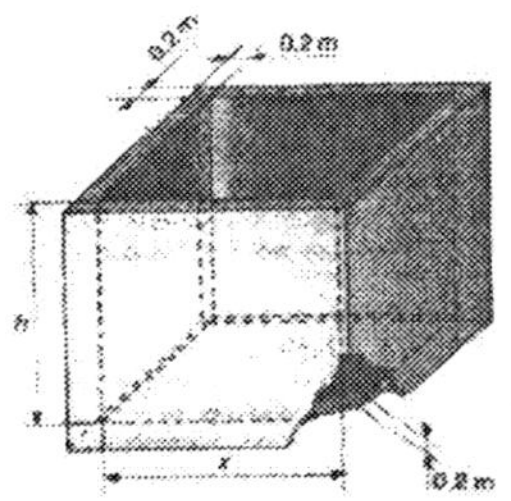

In Chapter 5 (§ 6), it was stressed that in the tank problem (Box 9-11), posed to students, which appears at first to be a problem of optimization corresponding to a very particular situation, the *motivation* for generalization results from peculiarities of the result found. The value of x which corresponds to the minimum is 2, a number connected to the data of the problem: this is for example half the number that measures in m^3 the internal volume of the tank. Is this a mere coincidence? The introduction of parameters to denote some of the data allows this question to be answered. If one denotes the thickness of the tank by e and reworks the computations with this parameter, one discovers that the value of x which corresponds to the minimum does not depend on e. Is the result obtained also independent of the chosen volume? This time, the answer is negative: if V is the internal volume (expressed in m^3), the minimum is obtained for $x \doteq \sqrt[3]{2V}$. A new functional dependence appears which can be studied...

What can we learn from the experiment about the potential of this problem for motivating and approaching generalization issues in a CAS environment, at high school level?

In this piece of engineering, this problem is posed to the students quite early, at a point when the computation of derivatives has not yet been well mastered. The issues of generalization which we choose to emphasize here are not the only ones justifying its didactic value at this point in the learning process and, before going on to these, we would like to point out some other potentials revealed by the experiment. The first problem met here is a *mathematization* problem. Some hints are given but to move forward, students must find a way of calculating the volume of concrete (to proceed by difference of volumes, which is obviously the most successful method here, is not a spontaneous strategy for many students). This volume, once obtained, depends on two variables x and h, and students have to eliminate one of these by using the relation $hx^2 = 4$ so as to be able to make use of their analytic tools. Even if this situation is *a priori* frequent at high school level when functions are related to the modeling of systems, generally, selecting the independent variable and expressing the others with respect to it is the responsibility of the teacher or the text of the exercise. Once these initial difficulties are overcome, students obtain for the volume V(x) the expression: $(x+0.4)^2(\frac{4}{x^2}+0.2)$ or an equivalent expression if they choose x as independent variable and some expression with square roots if they choose h. In the experiments, this phase of mathematization proved to be difficult. Students were not able to succeed without additional guidance, even though the teachers tried to give them maximal responsibility.

Students meet optimization problems from grade 10 onwards in France, but their culture for solving such problems is an algebraic-graphic one, not bringing in the notion of derivative. Under these circumstances, the scenario asked students, following a collective synthesis, to investigate the variation of the volume with respect to x and to make conjectures. This graphical exploration mediated by the TI-92 led to a conjecture that the function seemed to have a minimum in the neighborhood of 2. By symbolic computation, it would have been possible to show algebraically that 2 is a minimum: by calculating and factorizing V(x) – V(2). This strategy had been used in other functional situations at the beginning of the year. But, after this phase of exploration which was followed by a second overview, the scenario directed the students to an analytic solution exploiting the newly introduced notion of derivative. As they were not yet very familiar with the computation of derivatives, students were asked to calculate V'(x) with paper and pencil, and then to check the result obtained with the machine. This choice is not neutral. The function V(x) is complex and, depending on the way in which it is expressed, the computation of the derivative is more or less laborious, more or less a source of errors. It proves especially difficult if the students choose to use the 'simplified' form given by the calculator (Figure 9-8). In the classroom this motivated an interesting discussion about *forms and differentiation*, taking into account the management of quotients, and the respective advantages of expanded and factorized polynomial forms. Checking the computation with the calculator raised equivalence problems, already described in relation to the first project: no less than five different forms generally arose, without counting the erroneous ones, all different from the derivative given by the TI-92. To conclude, the study of the sign of the derivative showed the advantages of factorized forms, the numerator being a polynomial of degree 4. As can be seen from these comments, independently of any perspective of generalization, this situation is already a rich and complex one for grade 11 students.

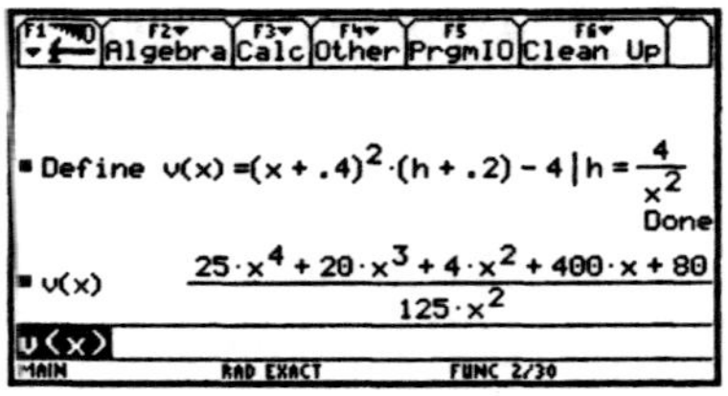

Figure 9-8. Expression for the volume on the TI-92

For the first experiment, generalization was introduced by posing the students two more questions to solve, presented as an extension of the first problem:

Extension (class B)

- solve the same problem with a thickness of the tank of e meters. (The volume of the tank is still 4 m^3);
- solve the same problem with a thickness of the tank of e meters and an interior volume of the tank of V_0.

The students tried to solve these two problems but they perceived neither their value, even *a posteriori*, nor the role played in generalization by the introduction of parameters. The lack of problematization was evident here. Once again, the first observations and the associated analyses led us to revise the scenario. The value of h, once found, had to be connected with the data so that generalization *made sense*. Tools available for such a generalization had to be collectively discussed so that the limits of numerical attempts be perceived and the introduction of parameters be considered as the appropriate mathematical tool. In fact, once these conditions were clarified, revising the scenario was quite easy and this revision turned out to be effective. As regards the technical work associated with generalization, this would have been impossible for these students without the support of the symbolic calculator. And, even if most of the students, for the first generalization, started the new computations from scratch, for the second one, they tried to save the computations: they came back to previous screens on the machine, adapted the expression for V(x) by introducing the second parameter, and then followed the steps of the process very economically.

The second problem belongs to the class of problems labeled as search for *functions satisfying certain constraints*[15], and was taken from the Belgian project AHA (1996)[16]. The didactic value of this kind of problem has been demonstrated by different researchers, for instance Rogalski (1990) and, more recently Bloch (2000, p.171). The problem was initially posed in the following form, for a 2 hour session:

Box 9-12.

A problem of pipes
(AHA 1996, p.47)

One wants to link the two pipes shown in section in the figure below.
The joint is generated by the rotation of a curve around the axis of the pipes. This curve must be tangent to both pipes at the junction points and its slope cannot change abruptly.
The purpose of the problem is to find such joints and to choose that or those which minimize the slope of the joint.

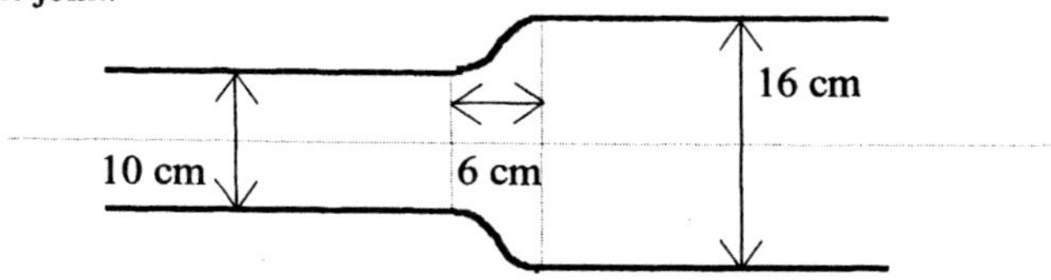

Generalization enters into this problem via the choice of one or several types of joints: by cubic functions or polynomial functions of odd degree, by sine functions, by combination of segments and arcs of circles, of parabolas... For every type, one deals with a family of functions depending on several parameters whose values will then be determined by expressing the constraints to be satisfied.

Even if grade 11 students have at their disposal, when this problem is set up, a rather rich repertoire of functions which allows them to enter into this type of generalization, the problem is doubtless difficult. They are not familiar with this class of problems, the constraints concern both the function and its derivative, not to mention the fact that the introduction of a functional frame is itself their responsibility. Furthermore, the problem does not stop with the determination of possible joints, they are asked to compare these by trying to optimize, subject to a general criterion concerning the derivatives. In order to shorten the text proposed, this criterion is formulated fuzzily, but what is aimed for is the minimization of the maximum value of the slope on the interval under consideration[17].

What responsibility can the students be given in solving such a complex problem and what are their needs in terms of instrumented knowledge? In the next paragraph, we try to draw lessons from the experiments carried out.

The lessons drawn from the first experiment

The first experiment proved that our students, if faced with this problem, were not, as a group, completely without resources. The fact that it is possible to make such a joint by rounding off a linear joint at its extremities was suggested immediately. Other students, quickly taking the functional point of view, envisaged different types of joints by cubics and sines, and by connecting two parabolas with their respective summits at the two extremities of the joint. It was their perspective which prevailed over the first one, more pragmatic, for evident reasons of the *didactic contract* (Box 8-6).

But the experiment also showed the *complexity* of the knowledge required by this mathematization and in solving this problem, once various possible joints are considered, as well as the problems raised by an effective instrumentation.

For most students, working with the TI-92 favored graphical investigations at first. These aimed at finding functions whose graph had the form of the joint, before adjusting the coefficients more precisely. After a phase of general discussion, they decided to begin with cubic functions. Thus they entered particular cubic functions into the calculator and graphed them, then changed the coefficients, trying to get a good form. The attempts were poorly structured and, as could be foreseen, the coefficients of the chosen cubics were small integers, so that their graphs did not take on the required form in the standard window used by the students, but were, on the contrary, very steep in form as a result of the high values taken by the terms of degree 3 (Figure 9-9). All the session could have been spent in these attempts!

Figure 9-9. 'Ordinary' cubics

Passing from the idea of using cubics to the idea of determining a cubic function within a family by choosing values for parameters, becoming aware that choosing these values cannot be done by chance but requires an adequate expression of the constraints to be satisfied, that expressing these constraints requires the choice of axis..., each of these steps required *interventions* from the teacher, as the strategies of those students who spontaneously worked in this way did not spread across the classroom. However, once the necessity of fixing coordinates was understood, most students were able to choose axes in order to take advantage of the symmetry of the situation. But, once these first difficulties were overcome, other new ones emerged linked to the expression of the constraint of smoothness. This constraint was interpreted by the students in the following correct way: the curve has to have horizontal tangent at the junction points, but many of these evidently had difficulty in reformulating this as saying that the value of the derivative is 0 at the junction points. Such difficulties resulting from a lack of *flexibility* between the different perspectives that can be adopted with respect to the derivative were omnipresent in the experiments. The time spent on cubic joints led the teacher to give up the initial idea of comparing the solutions according to the choice of the axis. And the collective summing up after this phase only considered the modeling in the system of coordinates predominantly chosen by the students.

The second type of joint envisaged in the class, involving two parabolas, can be dealt with in several ways. In continuity with what had preceded, students introduced general expressions depending on three parameters and tried to determine the parameters by expressing the constraints. The problem is less simple than it appears at first sight, if one does not reason in terms of the symmetry with respect to the point C in the middle of the segment linking the two extremities of the joint, even if one reasonably decides to connect the two parabolas at C. Indeed, the condition of tangency at C adds a new condition which involves the two functions, but it is a condition which is actually automatically satisfied as soon as the others are. One can also solve the problem more geometrically by considering a first parabola passing through C, then by considering its image by point-wise symmetry with respect to C. If one admits that symmetries preserve tangency, which seems to be accepted by these students, as observed several time in the experiments, the problem of connection at C is thus solved because the tangent to the first parabola is invariant in this symmetry. The problem of connection at the other extremity of the joint can then be managed in a similar way. The teacher outlined this possibility very briefly in the collective summing up.

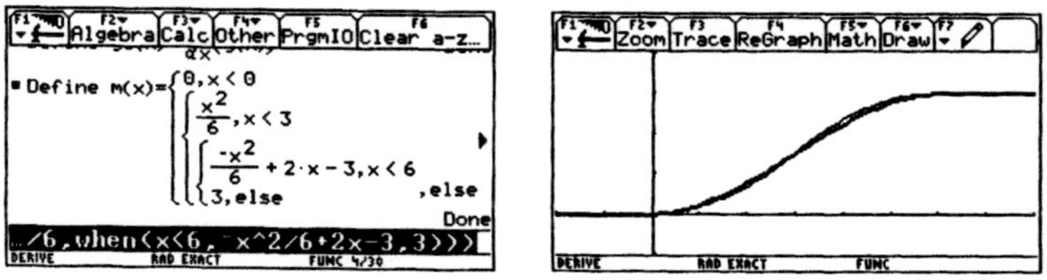

Figure 9-10. Expression of the parabolic joint and drawings of the cubic and parabolic joints (Artigue & al 1998, p.189)

The comparison of the first two joints found was then made without difficulty. The students calculated with the machine then drew the derivatives of both functions, which led them to conjecture that the maximum of the slope was obtained in both cases at the middle of the joint. The proof was not difficult because they had either a parabola (cubic joint) or two segments (parabolic joint). The comparison of the two maximum values: 3/4 and 1, showed that the cubic joint was better.

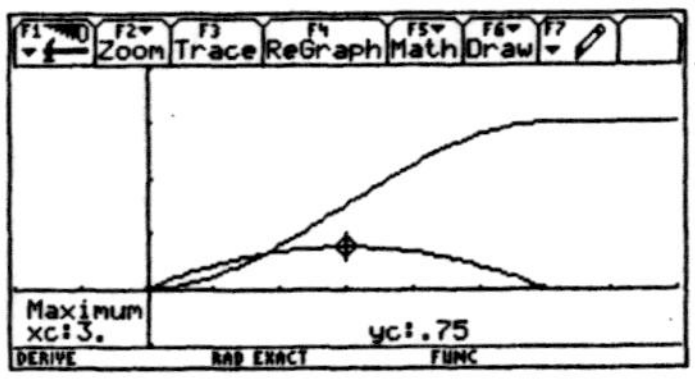

Figure 9-11. Drawing of the cubic joint and of its derivative (Artigue & al 1998, p.188)

The third type of joint considered at the end of session could also be managed via the writing of a general form depending on four parameters: f(x) = a.sin (bx + c) + d. But the equations obtained are unmanageable for a grade 11 student. In fact, for this third type of joint, reasoning by successive transformations, taking into account the characteristics of the targeted sine function and the existing links between geometrical and algebraic transformations, is doubtless the most economic approach for these students[18]. The teacher explicitly engaged the students in that direction at the end of the session, when she asked them to study at home the case of the sine joint for the next session.

This problem is thus a very rich one, in terms the types of functions it brings into play, and by the different points of view it requires to be introduced and connected, both within analysis and in interaction with geometry. For the students involved in this experiment, nothing was routine, every phase of the solving process involved something problematic. Contrary to the situation of the tank problem as previously described, calculators *did not necessarily play a positive role.* At the beginning of the session, the extensive use of graphical investigations, poorly structured and controlled, was not effective at all. Conversely, at the end of the session, graphical exploration turned out to be very useful when students looked for relations between the coefficients of sine functions and their graphical characteristics, or when they tried to compare the slopes of different joints. But what was at stake here, was essentially the graphic application. The symbolic application could have been brought into use for the solution of the linear systems involved, but the level of students' instrumentation as regards this type of task was rather limited, so they preferred to solve these systems by paper-and-pencil techniques.

Under these conditions, within the reduced amount of time devoted to this problem, what was observed was a functioning where the mediations of the teacher were *very important*. That is the reason why the same type of problem was then adopted in another experimental mechanism involving Internet exchanges between classes. This new mechanism which constitutes a particular orchestration (Chapter 8, § 2) allowed the research to develop over a longer period (approximately one month), and the students to benefit from the ideas of other classes and from competition between them. Similarly, one more situation (Trouche 1996, p.225) of the same category -- looking for functions subject to a set of constraints -- was used in a *piece of long term engineering* including research on open problems, piloted by the teacher but mainly developed in out of school time (Aldon 1999, p.170). This type of management makes it more easy to manage the *reorganizations* of knowledge that these situations require.

5. GENERAL DISCUSSION

The document accompanying the new grade 10 French syllabus (DESCO 2001) mentions:

> Recent developments in calculators allow us to anticipate everyday access to systems for formal computation. These have to be efficiently integrated into the whole learning process. Complemented by the necessary training for the mastery of the most common computations, they can represent a valuable means for checking, but also for investigating, opening the way to richer mathematical situations (p.10).

As pointed out at the beginning of this chapter, the examples of engineering presented here are guided by this perspective. They envisage an integration process in which the instrumentation and instrumentalization of the calculator on the one hand, the mathematical knowledge on the other hand, simultaneously develop through a dialectic which tightly connects paper-and-pencil work and work instrumented by the calculator.

The descriptions and analyses of this chapter tend to show that, at least in the experimental environments concerned here, *such an integration is possible*. They also tend to show that such an integration is likely to help high school mathematics education:

- to *bring alive* important but difficult questions whose management in standard environments is highly problematic, such as those relating to the *sense* of algebraic, expressions and objects, to the *relationships* between exact and approximate computations, to the *status* of numbers;
- to benefit better from the current visual culture in order to *motivate* the introduction of new concepts and the associated work of *mathematization*;
- to allow approaches to *generalization*, to *complexity*, at a point when students' competences make these approaches difficult in paper-and-pencil or even graphic calculator environments.

But the descriptions and the analyses of this chapter also show that such an *effective integration is not easy at all* and they help us to identify *conditions to be satisfied*. Some of these conditions were obviously achieved from the start of the engineering work described: the *legitimacy* of the symbolic calculators, inside the classroom, inside the high school, in the relationships with parents; the fact that even if the calculator was always available, its use in mathematical work remained under the control of the teacher and that the rules established in the class as regards its use were respected. The teachers involved in these experiments were also aware of the fact that these rules would necessarily evolve, throughout the year, according to the development of students' instrumental and mathematical knowledge, and that this evolution was only one facet of the necessary evolution of the didactic contract. They managed this evolution, making precise its most crucial points, in order to avoid possible misunderstandings. This management can lead to specific didactic inventions, such as the different codes introduced by the Montpellier team to assessment texts in order to make clear the status of the calculator in solving different questions as well as the level of recording expected: paper-and-pencil type solution, solution possibly referring to the calculator but mentioning the commands used with the corresponding syntax and the outputs obtained... The teachers were also aware of the *role* that they had to play in the instrumentalization and the instrumentation of the calculator, and they took up these processes officially from the start. However, in spite of the preparatory reflection carried out, the first didactic scenarios were far from being satisfactory and required serious reorganization. For instance, they did not pay enough attention to the fact that exploring a perceptual phenomenon via the screen of the calculator does not automatically generate real mathematical work; they tended systematically to *minimize the complexity* of the work which remains for the student when s/he works with a symbolic calculator; they tended to systematically *underestimate the mathematical needs* of instrumented work.

How can these recurrent phenomena be explained? In order to deepen our reflection, we will use the thesis study which Defouad (2000) conducted in the context of the second engineering project referred to above, addressed to grade 11 students. This thesis focused on understanding a particular instrumental genesis: that associated with the study of the variation of functions. Through regular classroom observations and the following up of a set of students, selected according to their sex and their mathematical and technological profile, throughout the academic year, this thesis showed the complexity of the instrumental genesis in question. Regular interviews were organized with the selected students and, among other questions, students were systematically asked to study a function they were not familiar with. With the help of the calculator, they had to explore the situation, propose conjectures about the behavior of the function, and then try to prove these. These interviews allowed Defouad to identify some important didactic phenomena which were not so visible in classroom sessions and assessments, due to the more constrained character of these.

A first phenomenon is the *slowness of instrumental genesis*. Defouad has shown that, facing an unfamiliar situation, in a context where the pressure of the didactic contract is lighter, the students rely longer than expected on a graphic culture of the study of functions, which has begun to develop in grade 10. This culture becomes richer as the students tend to progressively make use both of the graph of the function and the graph of its derivative in their explorations, conjectures and proofs, but the symbolic computations offered by the calculator and the modes of reasoning they support are, for a long time, confined to a marginal role. Typically, the symbolic application is used for calculating the derivative before entering it into the *Y*= application. The connections between symbolic, numerical and graphical approaches are thus poorer than expected. The economic strategies for use of the TI-92 envisaged *a priori* before every interview are rarely those chosen by the students who seem more likely to favor strategies of *zapping* and of *over-checking*. Moreover, this slow experimental genesis is marked by an alternation of phases that Defouad qualifies as phases of *explosion* and *purification*. The first ones are characterized by an explosion of different techniques and it is only slowly that a *temporary stabilization* takes place around schemes which can remain variable from one student to another.

These results lead one to wonder about the *status* of instrumented techniques in the experimental classes and about the influence of this status on the genesis observed. This question of the status of instrumented techniques along with that of their *institutional management* will structure the analysis we will develop below, by approaching, from a slightly different perspective, the data already analyzed above.

Education, in an environment of symbolic computation, inevitably links two types of technique: paper-and-pencil techniques and instrumented techniques. As has already been mentioned in Box 5-1, every technique has a pragmatic value and an epistemic value, the second value interesting education as much if not more than the first one. The *institutional status* of techniques depends on these values, and a technique which is not recognized as having sufficient epistemic value has difficulties in becoming fully legitimate. As has also been explained in Chapter 5 (§ 5), while it is easy to recognize a pragmatic value in instrumented techniques, it is much more difficult to grasp their possible epistemic value. In some sense, while the epistemic value of a paper-and-pencil technique often emerges from the details of the technical gestures attached to its execution, the immediacy with which results are obtained makes *the possible epistemic value of instrumented techniques less evident*: making visible this possible epistemic value requires reflection and some reorganization of the tasks proposed to students. We see this reflection and reorganization as something *essential for legitimizing* symbolic tools in education, and for overcoming the evident limitations of the institutional discourse in that respect. If we look back at the two examples of engineering analyzed in this chapter, we can see that different levers have been used for that purpose, consciously or not:

- the *surprise lever*: this plays on the effect of surprise produced by unexpected results so as to destabilize erroneous conceptions, to promote questioning, to motivate mathematical work;
- the *multiplicity lever*: this plays on the potential offered by technology for producing a great number of results very quickly, so as to promote the search for regularities and invariants and to motivate mathematical work aiming to understanding these;
- the *dynamic lever*: this plays on the dynamic potential of graphic representations to overcome the evident limitations of paper-and-pencil work and to promote a dynamical way of approaching mathematical concepts and problems, the potential of which is now fully acknowledged.

But, even if these levers can be identified and if one decides to make use of them, their *effective exploitation* is not obvious. Our observations, like those carried out by various other researchers, for example Schneider (Appendix 4-1), illustrate this very well. In order to better understand the difficulties confronting an effective exploitation, it is necessary to deepen our reflection on the institutional status of the instrumented techniques and their relationship with paper-and-pencil techniques.

What Defouad's thesis shows, is the difficulty experienced by teachers, even in the experimental environments, in giving an *adequate status* to instrumented techniques and to *managing them institutionally*. In fact, the observations made during the first year of the research showed that paper-and-pencil techniques and instrumented techniques had very different lives within the class.

After a first classic phase of investigation and craft work, some paper-and-pencil techniques for the study of variation became *official*. They became the object of specific training and some of these even became *routines* (Box 5-1). A more *theoretical discourse* goes alongside their implementation, having explanatory and justificatory aims, even if fundamental theorems such as the theorem linking the sign of the derivative on an interval with the monotonic variation of the function are not formally proved at this level of schooling.

The situation was not the same for the instrumented techniques involved in the study of variation: techniques for the framing of graphic representations, techniques associated with the determination of the sign of the derivative, techniques allowing recognition of the equivalence of algebraic expressions... The calculator *expands the range of possible actions* for solving the corresponding tasks: consider for instance the number of types of zoom offered in the graphic application, the different commands which can be used for determining the sign of an expression by solving equations, factoring, and also the different commands which can be used for testing equivalence, both in exact or approximate mode. Teachers faced this multiplicity and had obvious difficulties in selecting instrumented techniques, which, supported by the culture, they would have done without effort in more standard environments. The spontaneous tendency seemed more towards showing the students the richness of the potential offered by the machine, which produced a technical congestion. As the institution did not give them rules, they were less sensitive to the necessity of choosing, and not equipped to do so. There was an *explosion of techniques* which thus remain in a relatively simply-crafted state, and this did not favor progression towards an effective instrumented activity.

Another difference can be added to the previous ones. Any technique, if it wants to be more than a simple gesture mechanically learnt, must be accompanied by a more theoretical discourse. For the official paper-and-pencil techniques, this discourse is known and can be found in textbooks or teacher material. For instrumented techniques, *it has to be built* and its elaboration raises specific difficulties as it necessarily brings both mathematical and technical knowledge into play. For example, mastering the instrumented techniques associated with the graphic representation of functions requires both mathematical knowledge about functions, about discretization processes and their possible effects and more specific knowledge about the way in which the artifact implements these discretizations (Box 6-2). Mastering the instrumented techniques associated with the treatment of algebraic expressions requires both algebraic knowledge and more specific knowledge about the symbolic representation of algebraic expressions in CAS, the notion of canonical form, the differentiation between semantic and syntactic equivalence. As has been stressed in this book (Chapters 2, 3 and 6), this specific knowledge *is not always easily accessible*, and can even be out of reach, when it concerns the algorithms implemented in the machine. For these reasons, building a theoretical discourse adapted both to the instrumented techniques concerned and to the level of students is not obvious. The observations we made during the first year of experimentation confirm this difficulty and its negative effects: instrumented techniques did not take on a real mathematical status and their epistemic value remained limited.

Finally, the last phenomenon which marks the differences between the two types of technique is a direct consequence of what has come before: instrumented techniques were not institutionalized in the same way as official paper-and-pencil techniques, even when they seemed fully legitimate in the class. This gave them an intermediate status and Defouad introduced the expression *locally official techniques* to label this status.

We were not sensitive to these differences at the beginning of our work and we did not analyze them immediately in these terms. It is the alternation of the phases of explosion and reduction, the slow and very personal stabilizations observed during the first experimental year that drew our attention to the particular life of instrumented techniques. The analyses we developed of the data collected then allowed us to identify the differences mentioned above and the difficulties which attested to these. This analysis had clear effects on the organization of the experiment during the second year. The teachers, made sensitive to these questions, tried to face up to them and we helped them to set up adapted explanations, justifications, institutionalizations for those instrumented techniques which they chose to privilege and to make official[19]. We also paid more attention to the *necessary evolution of the didactic contract* as regards instrumented techniques, according to the advance of mathematical and instrumental knowledge. This resulted in a very positive evolution of the instrumental genesis for most students, both in time and quality.

The viability of an integration of symbolic calculators in mathematics education requires that a real institutional status be given to some *coherent and substantial set* of instrumented techniques. This supposes choices between those which are possible *a priori*, as well as the development of a discourse allowing these techniques to be explained and justified, a discourse supporting their institutionalization. This also supposes that the mathematical needs of instrumented work be recognized, and that it be accepted that, for a given mathematical domain, these needs do not coincide with the needs of paper-and-pencil work.

This is not obvious and requires, beyond official discourses and texts, an adequate training of teachers. Our experience with pre-service teachers preparing for the CAPES examination (note 5, this chapter) or having just passed it, tend to show that their initial training does not equip them well, mathematically and didactically, to take care of these problems. Research carried out on in-service teacher training concerning ICT shows that training sessions in this domain have limited efficiency. Often trapped in some kind of militancy, having as its first aim to sweep away the teachers' resistance, it does not provide them with means for overcoming the difficulties they necessarily meet. As shown by Abboud in her thesis (1994) teacher trainers are generally teachers expert in the use of ICT; they are no longer conscious of all that they had to learn in order to gain this expertise. Training is based on *imitation*: trainers propose teaching situations they have built and used with their students; they select those that best demonstrate the interest and innovative power of technology for mathematics teaching, but unfortunately these often also require much technological expertise from the teacher.

During the training sessions, these situations are simulated, the teachers playing the role of hypothetical students, according to the strategy of *homology*[20] so frequent in teacher training. They are then invited to use some of the simulated situations with their own students and, in the best cases, are asked to report on these experiences in further training sessions and are thus offered the opportunity of comparing these experiences with that of their colleagues, and of discussing these with the trainer. Taking account of the analysis developed in this chapter, and more generally in this book, such strategies, in our opinion, are *inadequate in facing effectively the complexity* of the problems involved. And it is not by chance that most teachers attending such training sessions do not take these further in their own classes. Other strategies must be developed. The chapter we dedicated to this issue in Tinsley & Johnson (1998) has already presented some alternatives. Others will be considered in the last chapter of this book.

In conclusion, we would like to underline that the didactical engineering presented in this chapter shows that symbolic tools can support the development of mathematical knowledge through different categories of situations and especially in the following two ways:

- through the *mastering of instrumented techniques*, which at first appeared rather as a constraint;
- through the *new potential* offered by instrumented work in symbolic environments.

Situations from the second category are easier to conceive because they often appear as enrichments of situations that already exist in customary environments. Situations from the first category, on the contrary, often have no immediate equivalent in customary environments and are, therefore, much less present in the literature, including the research literature. However, they cannot be neglected because, as has been pointed out in this chapter, they contribute in a essential way to the epistemic status of instrumented techniques.

In our opinion, situations organized around problems of equivalence and simplification, approached here in the grade 10 example of engineering, are paradigmatic of the first category whereas situations giving access to generalization via symbolic computations involving parameters are paradigmatic of the second category, for secondary education at least. In the first case, knowledge comes from reflection on the *object* (Box 9-1); in the second case it is the *tool* which is used through the potential it offers for quickly obtaining numerous results, for accessing the details of a computation already done and for adapting it to the case where one item of numerical data becomes a parameter, for using the software as an assistant for computation and proof.

Understanding the potential of symbolic tools for the learning and the teaching of mathematics requires a deep reflection on the potential epistemic value of instrumented techniques, where these two facets so different *a priori* are simultaneously taken into account. Such a reflection, of course, should not be thought of as something absolute but is something which depends on context, both cognitive and institutional. From the clarification of potential to the achievement of effectiveness is a long distance to cover: designing adequate situations and more general progressions, testing their viability and robustness, by taking into account the connection and complementarity between paper-and-pencil techniques and instrumented techniques, as well as the necessary *institutional negotiation* of certain mathematical needs, a negotiation which *is not so easy* today.

Michèle Artigue (artigue@math.jussieu.fr) is a professor of mathematics at the University Paris VII Denis Diderot, director of the IREM Paris VII and researcher in mathematics education. During the last ten years, her research in this area has mainly dealt with the integration of computer technologies into secondary mathematics education and with the didactics of elementary analysis or calculus. She is also vice-president of the International Commission on Mathematical Instruction.

NOTES

1. Problematics: this term refers to the structured set of questions addressed in research.
2. This is shown by the recurrent debates around mathematics education and leads to a distinction between what is a matter of social utility and what has more a cultural value. The recent book *Mathematics and Democracy*, produced by the National Council on Education and the Disciplines, in the USA (NCED 2001), makes a distinction between mathematical learning and the *quantitative literacy* that is today necessary for citizens living in democratic countries. The authors show that quantitative literacy requires much more than is traditionally taught about numbers and data processing, in the contemporary world where the critical management of numerical data, probabilistic and statistical reasoning play a crucial role. They also point out that the competences required by quantitative literacy cannot be considered as some natural by-product of a classical mathematical education.
3. Through a historical analysis of educational practices over the twentieth century, Birebent (2001) shows in his thesis that this problem with the checking of approximations already existed before the introduction of calculators. The introduction of scientific calculators certainly modified practices in this domain, introducing new routines and, for example, the idea that, in normal computations, it is enough to calculate with the maximal precision of the machine then to drop two or three decimals so as to get an estimate of the corresponding precision. By taking the example of simple trigonometrical computations, Birebent shows how this naïve idea does not take into account the properties of the functions involved in the computation.
4. The reader can consult for instance the report on computation produced by the CREM (Kahane 2002).
5. CAPES: the competition in France in which success gives access to a position as a teacher in secondary education. It is open to students who have gained the "Licence" (a university degree awarded after three years of study, broadly equivalent to a bachelor's degree in the US and the UK).
6. This distinction, initially introduced by Frege (1892), between sense and denotation has been used by various educational researchers to analyze students' functioning and to understand the difficulties they meet in algebra. The reader can refer for example to (Arzarello & al 2001) or to the research developed by the team GECO in France. This team exploits these ideas to define strategies for remediation for high school students (GECO 1997).
7. MAPLE for instance offers 20 options for the command *Simplify*. These make it possible to use some rules of simplification while ignoring others.
8. A polynomial $P(X) = a_0 + a_1X + \ldots a_nX^n$ can be written in the following form: $a_0 + X(a_1+X(\ldots+X(a_{n-1}+Xa_n)))$. This form, known as the Horner's form, is better adapted to the computation of numerical values of P(X) as the computation requires n products instead of $1 + 2 + \ldots n - 1$ products for the usual form.
9. As soon as the factorization of an expression involves square roots, one needs to use the command *Factor(expr, var)*.
10. The part of this piece of engineering concerning the equivalence of algebraic expressions was used with grade 11 students in the trial carried out by the research team DIDIREM (Artigue & al 1998). In this trial, already described in (Box 5-6), the greater mathematical maturity of the students made predictions, interpretations and checking easier to engage with and manage, and nearly the same design led to very interesting and useful mathematical work.

11. In these sessions, secondary mathematics teachers present the functionalities of calculators and examples of their use in classrooms (Chapter 1, § 3.5.2).
12. This document, entitled *Introduction au calcul formel*, is one of the documents distributed by Texas Instruments within the framework of T^3 Europe (Chapter 1, § 3.5.2).
13. The expression: *reference function* is used in the high school syllabus in France. It refers to the functions: x^2, x^3, x^n, $1/x$, $\sqrt{x}$ and $|x|$. These *prototypical* functions are first introduced and then teaching pays specific attention to the way the properties of functions from the associated families, called *associated functions* can be obtained by using geometrical transformations which make it possible to pass from the graph of a function to the graph of the reference function from the same family (for instance $(x - 1)^2 + 4$ is associated with x^2 and $|x - a|$ with $|x|$) (Artigue 1993).
14. * in Box 9-10 refers to the fact that, in the previous phase some students began to compute the derivatives corresponding to Box 9-9.
15. Another example is given by Aldon (1995, p.31).
16. The group AHA: *Approche Heuristique de l'Analyse* has developed a project for the teaching of analysis in the two last years of Belgian high schools relying on the results of didactic research in that area.
17. An optimal solution to this problem does not exist because one can build joints for which the maximum of the slope is as close as one wants to the limit value given by an affine joint (which does not answer the problem).
18. Working on these links and the way they make it possible to pass from one function from a particular class (second grade polynomial functions, homographic functions, for instance) to another one in the same class is an explicit aim of the French syllabus from grade 10 onwards.
19. These choices could depend on the possibilities offered at this level.
20. This term was introduced by Kuzniak in his doctoral thesis (1994) for labeling training strategies where the trainer, putting the teachers in training into the position of a student, uses teaching situations and methods with them which he wants the trainees then to reproduce with their own students. Kuzniak shows the present dominance of these strategies in the training of elementary teachers.

REFERENCES

Abboud-Blanchard M. (1994) *L'intégration de l'outil informatique à l'enseignement secondaire des mathématiques: symptômes d'un malaise,* Doctoral thesis. Paris: Université Paris 7.

AHA (1996) Une approche heuristique de l'analyse, *Repères-IREM* 25. Metz: Topiques Editions, 16-28.

Aldon G. (1995) Une voiture à la DERIVE, *Repères-IREM* 21, 27-44.

Aldon G. (1999) Démarche scientifique et calculatrices symboliques, in D. Guin (Ed.), *Calculatrices symboliques et géométriques dans l'enseignement des mathématiques* (pp. 165-175). Montpellier: IREM, Université Montpellier II.

Artigue M. (1993) Enseignement de l'analyse et fonctions de références, *Repères-IREM* 11, 115-139.

Artigue M., Defouad B., Duperrier M., Juge G. & Lagrange J.B. (1998) Intégration de calculatrices complexes à l'enseignement des mathématiques, *Cahier DIDIREM n° special 3*. Paris: IREM. Université Paris 7.

Artigue M. & Lagrange J.B. (1999) Instrumentation et écologie didactique de calculatrices complexes: éléments d'analyse à partir d'une expérimentation en classe de première S, in D. Guin (Ed.) *Calculatrices symboliques et géométriques dans l'enseignement des mathématiques* (pp. 15-38). Montpellier: IREM, Université Montpellier II.

Arzarello F., Bazzini L. & Chiappini G. (2001) A Model for Analysing Algebraic Processes of Thinking, in R. Sutherland & al (Eds.), *Perspectives on School Algebra.* Dordrecht: Kluwer Academic Publishers, 61-81.

Birebent A. (2001) *Articulation entre la calculatrice et l'approximation décimale dans les calculs numériques de l'enseignement secondaire français,* Doctoral thesis. Grenoble: Université Grenoble I.

Bloch I. (2000) *L'enseignement de l'analyse à la charnière lycée/université. Savoirs, connaissances et conditions relatives à la validation*, Doctoral thesis. Bordeaux: Université Bordeaux I.

Bronner A. (1997) *Etude didactique des nombres réels, idécimalité et racines carrée,* Doctoral thesis. Grenoble: Université Grenoble I.

Brousseau G. (1997) *Theory of didactic situations*. Dordrecht: Kluwer Academic Publishers.

Defouad B. (2000) *Etude de genèses instrumentales liées à l'utilisation de calculatrices symboliques en classe de première S,* Doctoral Thesis. Paris: Université Paris 7.

DESCO (Ed.) (2001) *Accompagnement des programmes - Mathématiques - Classe de première des séries générales*. Paris: Centre National de Documentation Pédagogique.

Douady R. (1986) Jeux de cadres et dialectique outil/objet, *Recherches en Didactique des Mathématiques* 7(2), 5-32.

Fortin P. (1998) Introduction au calcul formel, *document Texas Instrument,* T3 Europe.

Frege G. (1892, 1971) *Ecrits logiques et philosophiques*, Traduction C. Imbert. Paris: Seuil.

GECO (1997) Comment recueillir des connaissances cachées en algèbre et qu'en faire? *Repères-IREM* 28, 37-68.

Guin D. & Delgoulet J. (1997) *Etude des modes d'appropriation de calculatrices graphiques et symboliques dans une classe de seconde*. Montpellier: IREM, Université Montpellier II.

Kahane J.P. (2002). *L'enseignement des sciences mathématiques.* Paris: Odile Jacob.

Kuzniak A. (1994) *Etude des stratégies de formation en mathématiques utilisées par les formateurs de maîtres du premier degré*, Doctoral Thesis. Paris: Université Paris VII.

Laborde C. (1999) Vers un usage banalisé de Cabri-Géomètre avec la TI-92 en classe de Seconde: analyse des facteurs de l'intégration, in D. Guin (Ed.), *Calculatrices symboliques et géométriques dans l'enseignement des mathématiques, Actes du colloque francophone européen* (pp. 79-94). Montpellier: IREM: Université Montpellier II.

Lakoff G. & Nuñez R. (2000) *Where Mathematics Comes From: How the Embodied Mind Brings Mathematics into Being*. New York: Basic Books.

National Council on Education and the Disciplines (Eds.) (2001) *Mathematics and Democracy. The Case for Quantitative Literacy,* The Woodrow Wilson National Fellowship Foundation.

Oldknow A. & Taylor R. (2000) *Teaching Mathematics with ICT*. London & New York: Continuum.

Rogalski M. (1990) Réhabiliter le graphique, *Audimath*. French Ministry of Education.

Tinsley D. & Johnson D.C. (Eds.) (1998) *Information and Communication Technologies in School Mathematics*. London: Chapman & Hall.

Trouche L. (1996) Enseigner les mathématiques en terminale scientifique avec des calculatrices graphiques et formelles. Montpellier: IREM, Université Montpellier II.

Trouche L. (1997) *A propos de l'enseignement des limites de fonctions dans un environnement calculatrice, étude des rapports entre processus de conceptualisation et processus d'instrumentation,* Doctoral Thesis. Montpellier: IREM, Université Montpellier II.

CONCLUSION

The contributors to this book set out with two main objectives: on the one hand, to seek reasons for the continuing marginality of symbolic calculators within mathematics teaching; on the other, to identify conditions necessary to make their integration *viable*.

We have not sought to minimize the problems confronting integration of symbolic calculators in particular, and of information and communication technology in general. On the contrary, we have wanted to provide readers -- whether they be policy-makers, researchers, or teachers -- with means -- fitting these various levels of involvement -- to advance such integration. We have also wished to provide means to analyze and evaluate pedagogical resources, and to make them more efficient, beyond the particular examples described in this text.

We have done this by developing a theoretical framework for thinking about innovative experiences of calculator integration, by finely analyzing such experiences, and by providing evidence of their successes and their limitations. We have sought to identify the processes which produced these successes and the conditions which would allow them to be *reproduced* more widely, and to show how certain limitations can be surmounted by giving attention to phenomena which the education system tends to hide or underestimate.

The necessity of studying the instruments constructed by students

We have shown how students have quite rapidly adopted, first, scientific calculators, then graphic, and finally symbolic, as tools for their mathematical work. Analysis of their usage has brought to light phenomena linked both to *computational transposition* and to individual processes of adaptation to tools. The affordances and constraints of these tools play a determining role in these processes of adaptation, but the observations reported here show that this role is not always positive as regards processes of conceptualization. The *instrumental approach,* which emphasizes the *instruments* constructed by students, makes it possible to understand these phenomena and to describe them precisely, distinguishing different types of constraint and their influence on instrumental genesis. Instrumented action can be described in terms of *schemes*, including *operational invariants*, relating the observable competences of students to their conceptions:

- an instrument consists of a part of the artifact and the schemes of instrumented action through which a student accomplishes a given task;

- the processes of instrumental genesis have both individual and social aspects;
- an instrumented technique, the observable part of a scheme, allows the action of a student to be described in terms of gestures;
- there is a dialectical relation between gestures and operational invariants, the operational invariants *guiding* gestures and gestures *instituting* operational invariants.

Studying instrumented techniques consists of studying their pragmatic, heuristic and epistemic functions. This calls for more precise analysis of students' actions and accompanying discourses over time, to identify regularities in the carrying out of tasks and the justifying of gestures.

Techniques, understood as interfaces between action and conceptualization

Thus, conceptions, instrumented techniques and 'customary' paper-and-pencil techniques are intimately connected. In a general way, techniques play a fundamental role in teaching and learning, in terms of *connecting action* and *conceptualization*. They are constitutive components in the understanding of mathematical objects, and sources of new questions. Taking account of instrumented techniques and of their interaction with customary techniques is, then, one key to the integration of new tools. This is not simple:

- it is sometimes necessary to defer the use of certain 'key-press' instrumented techniques so as not to deprive students of techniques involving exploration which play an important role in conceptualization;
- an instrumented technique can be *described* and taught. However, the studies reported here have shown that a technique taught does not necessarily become a technique adopted;
- the introduction of new instrumented techniques depends on the development of new *praxeologies*, starting from new tasks requiring use of these new techniques;
- use of these techniques by students depends on their integration into the strategies of the teacher, in terms of the tasks set for students and the *discourse* used to present and justify corresponding techniques.

Giving *institutional status* to a coherent set of techniques, and providing a theoretical discourse accompanying these techniques are necessary to develop a *reasoned instrumentation* through using computational tools in the classroom, coordinated with the systems of instruments already established. Moreover, many examples presented in this book show how an instrumented practice which is efficient and well regulated depends on substantial mathematical knowledge. Nevertheless, these new equilibria are not easy to establish, for at least two reasons:

- CAS considerably extend the field of possible actions for accomplishing a given task (if one thinks, for example, of the number of zooms offered by the graphing application, or of the number of commands available to solve an equation). From this flows an *explosion* of *techniques*, which makes much more difficult the usual cycle of discovering, making official and routinizing techniques;
- establishing theoretical discourses to accompany techniques is often difficult when they depend on knowledge which is not easily accessible, linked to the programs implemented in the machine.

A new kind of relationship to mathematics

In the light of the work described in this book, we can bring out further points which need to be taken into account in creating these new praxeologies (whether or not they are accompanied by some evolution in mathematical content). Even if programs of study and official guidance make more and more frequent reference to tools, there is a clear lack of attention within the institution of schooling to conditions for their *viability*. An ecological approach provides some insight into these didactical phenomena, capable of guiding changes linked to new environments. Thus, the integration of CAS depends on creating a new type of relationship to mathematics on the part of teachers and students:

- in the past, the educational legitimacy of mathematics has been, at least in the scientific streams of secondary schools, as much (if not more) epistemic than pragmatic (since the teaching of mathematics aims more that students learn to *understand* than to *do*). The legitimation of CAS assumes a different balance between these, justifying the learning of an efficient instrumented practice of mathematical computation;
- in the past, the teaching of mathematics has valorized reasoning over computation, algebraic computation over numeric, exact numeric calculation over approximate. The introduction of CAS makes possible, but also requires, a new equilibrium to be found;
- the introduction of CAS makes it possible to bring to life mathematical questions which are important but difficult to manage such as those concerning the status of numbers or the sense of algebraic expressions;
- finally, the introduction of CAS depends on establishing new relationships between work involving a localization of attention (such as the use of zooms or the observation of particular algebraic expressions) and work involving *critical reflection* and generalization from particular conditions.

These new relationships have already made an appearance in studies reporting the integration of scientific calculators; they allow *intelligent calculation* to be developed by students. In the same spirit, new programs of study in primary schools place an emphasis on the coordination needed between mental calculation, knowledge of the properties of mathematical objects, and use of the tool for calculation. At another level, with symbolic calculators, this intelligence can develop through coordinating elements of algorithmics and programming on the one hand, and the use of preprogrammed commands on the other. With 'basic' scientific calculators, as with symbolic calculators, this is a matter of contributing to a reasoned instrumentation of ever more complex tools.

New curricula

Other factors play a part in explaining the marginal integration of computational tools, in particular CAS, within mathematics teaching. They concern the organization of curricula, from the point of view of the material conditions of teaching, and of the evolution of programs of study and official examinations:

- the *equipping* of schools, and the *accessibility* of machines and software to students and teachers, are, of course, conditions necessary for integration into teaching practices (and equipment and accessibility are, at the moment, very inadequate);
- the creation of *mathematical laboratories* (CREM report: Kahane 2002) would be quite useful from this point of view (although doubtless this would require work to convince mathematics teachers, who generally consider that the teaching of their subject does not contain an experimental dimension, in contrast to the physical sciences:
- it is necessary to introduce explicit objectives into programs of study dealing with the *knowledge* (including *know-how)* likely to promote the development of an algorithmic spirit on the part of students and to put conditions in place for an experimental approach. New forms of working, allowing groups of students to carry out interdisciplinary projects on a given theme could provide opportunities to develop such an approach;
- this integration needs to be accompanied by establishing new problems and new methods of solution which put this knowledge and know-how to work; Chevallard (1992) underlines the importance in this respect of didactic exploitation scenarios involving the *simulation* of systems or processes;
- it is necessary to incorporate the *time* necessary for the didactical management of these objectives, because experience shows that modification of praxeologies through contact with computational tools does not necessarily result in a saving of time, contrary to a widespread way of

thinking: experimental approach, instrumented action, and theoretical elaboration, through the formulation of hypotheses, can only be organized over *long-term* study of a domain;

- finally, this integration requires assessments containing items which imply a more experimental approach to more open problems. Their solution must call for the *use of instrumented techniques* and of a reasoned instrumentation of artifacts. In this way, one could test, for example, the capacity of students to interpret a response or non-response from an artifact, to make a conjecture, to evaluate the relevance of a tool to a given situation, to coordinate different applications of a CAS in order to validate an answer.

It has to be said that, at present, the institution of schooling has not yet integrated into its examinations the kind of evolution that it specifies in its programs of study. However, assessment plays an important role as a motor for the development of professional practice; the proposals above aim to encourage this. From this point of view, the frequent banning of calculators in the early years of university study (whereas they are permitted at upper-secondary level, and form a prescribed part of teacher training) creates a most unfortunate break.

The didactical implementation of CAS

We have shown through numerous examples how reasoned instrumentation with these tools requires a basic knowledge on the part of the teacher of underlying programming structures and of the algorithms employed, and more generally, of the constraints and affordances of these artifacts. The next stage consists of conceiving situations which exploit these constraints and affordances and which aim to develop students' reasoned instrumentation in the course of the process of conceptualization. In effect, numerous observations of the behavior of students, as described in this book, provide evidence that this reasoned instrumentation *does not arise of itself* and that situations must be thought out in order to support its development. We have also provided evidence of a differentiation of instrumental genesis in symbolic calculator environments: the more complex the tool, the greater this differentiation seems to be. Thus, some guidance of instrumental genesis is essential; it requires didactical management of the time and space of study. Such reflection on *didactical engineering*, relative to the objectives of teaching, considers, on the one hand, the *didactical exploitation scenarios* proposed for the situations devised in conjunction with the *instrumental orchestrations* underpinning each situation, to specify the *didactic configuration* and the ways in which it can be exploited so as to orient the instrumented action of students and the construction of systems of instruments.

New mechanisms for the training of teachers

This book has brought out the fact that computational tools are rarely considered by teachers as tools for mathematical work, in contrast to the reactions found amongst students. Chapter 4 has also shown the great variation in behavior amongst teachers. It seems as if, just as for students, the more complex the environment, the greater the *diversity of behavior across teachers*. This situation is easily explained by the competences required on the part of the teacher: knowledge of mathematics and computing necessary for an efficient instrumented practice, knowledge of the constraints and affordances of these artifacts, the implementation of new situations and, more specifically, of pieces of didactical engineering and corresponding instrumental orchestrations. This implementation involves a *profound questioning* of professional practices which, as shown, are linked to the conceptions which teachers have of mathematics. Such an evolution cannot be expected within a framework of continuing professional development limited to courses lasting a few days. Consequently, the integration of CAS, and more generally of other forms of ICT, calls for new mechanisms of professional development which provide *continuous* long-term support for teachers in their efforts at integration and which help them make the critical transition to pedagogical action.

In the development of these mechanisms, the idea of conceiving *usage scenarios* has proved particularly relevant; this idea acknowledges the necessity of taking account of the pedagogical organization of a class and the role of the teacher, as manifested in analysis of the logic of teacher-tutor cooperation in using intelligent tutoring systems (Vivet 1991). Such an approach aims at creating learning conditions in which technological environments can provide spaces for discovery with flexible tutorial assistance. The idea of building an *evolving network* of teachers to develop usage scenarios for geometry software was introduced in the USA even before the means fully existed (Allen & al 1994, 1996).

More recently, *usage scenarios* have been developed for teachers wanting to produce teaching units integrating the version of Cabri-Geometry implemented on the TI-92 calculator. These scenarios consist of the presentation of a unit with its objectives, student materials, and supporting notes for teachers to help put the unit into practice (Laborde 1999). A similar training mechanism has been developed around units integrating a lesson presentation, a usage scenario, and accounts of classroom explorations of these units by teachers in training, aiming both to assist management of the unit by the teacher and to promote *collaborative* work in the class around a scientific debate (Guin, Delouget & Salles 2000). This approach to organization has been extended through employing a distance platform to conduct collaborative workshops aimed at devising pedagogical resources, and to provide *continuous long-term* support for integration by teachers (Guin, Joab & Trouche 2003). Such pedagogical resources can be thought of as instruments evolving within a *community of practice* (Wenger 1998). The structure of these pedagogical resources was devised with the aim of facilitating both their implementation in classrooms and their evolution in response to teachers' ideas and experiences. Providing the resources are not too complex, it appears that this mechanism may help teachers to make the transition to pedagogical action.

The necessity of multidisciplinary research within the framework of the cognitive sciences

Integration of technological tools into teaching calls for a *chain of technical solidarities* (Chevallard 1992), from programmer to teacher by way of didactical engineer. This integration, and the collaborative work that it assumes, constitutes a radical change in professional practice. Thus, deeper reflection is needed on forms of working and communication likely to support such an evolution. This requires multidisciplinary research within the framework of the cognitive sciences: psychology and cognitive ergonomics, communication, computing and informatics, sociology, and subject didactics are all disciplines bearing on these problems. All the research perspectives outlined above are important for the educational community in relation to situations promoting reasoned instrumentation with CAS, and the exploitation of these scenarios through didactical engineering of instrumental orchestrations and associated usage scenarios.

The necessity of further exploration

In this work we have shown the difficulties but also the potentialities of symbolic calculators and, more widely, of computer algebra systems in the teaching of mathematics. At one level, of course, these studies relate to particular explorations, employing specific technologies, carried out in specific contexts, but they provide pointers to potentially important results of much wider significance.

We would encourage a broadening, in several ways, of the empirical base of studies of instrumentation. First, we welcome the appearance of studies which use a conceptual framework drawing on ideas of instrumentation to examine the role of dynamic geometry systems (DGS), particularly in relation to the teaching of geometrical exploration and argumentation (Laborde 2003; Mariotti 2002); these provide a very useful form of contrast to studies examining the role of CAS in algebra and calculus. Equally, of course, many other tools lend themselves to analysis in these terms, again providing further valuable contrasts for refining instrumentation theory.

Second, employing the language (and conceptual framework) of Dutch realistic mathematics education, there would be great value, as Drijvers & Gravemeijer suggest (Chapter 7), in studies which explicitly address classroom situations in which the emphasis is on horizontal mathematization in contrast to vertical, and with the aim of supporting *progressive mathematization* rather than a more direct diffusion of canonical mathematics. Switching to the language of French didactique, this might also contribute to a fuller and tighter analysis, across a range of studies, of how issues of instrumentation interact with the design of situations for action, formulation, validation and institutionalization.

This relates to a further aspect of broadening. All but one of the studies reported in this book have taken place wholly at secondary-school level, in what may be (in some respects) distinctive institutional circumstances, connected, for example, to regularities in the organization and ethos of such schools. Earlier work (Ruthven & Chaplin 1997; Ruthven 2001) suggests that developing a reasoned instrumentation of calculators within primary-school mathematics may display both similarities and contrasts to the secondary-school; the contrasts related (perhaps most directly) to a differing curriculum rationale (emphasizing, for example, the making of connections with students' everyday and wider experience, and the building of broad flexible competences), but also (more indirectly) to the rather different backgrounds and identities of primary-school teachers. Perhaps, too, moving beyond compulsory schooling, contrasting institutional circumstances mean that issues of instrumentation play out rather differently in, say, technical-vocational courses (with an emphasis on pragmatic over epistemic issues, and stronger contextualization in specific workplace practices), compared to university-scholarly mathematics courses (with locally framed curriculum and assessment, and perhaps a stronger emphasis on formal/structural/symbolic modes of argument over informal/analogic/graphic).

Finally, there is the intriguing question of what happens as teachers and students gain access to machines offering not just multiple functionalities, but real possibilities of interplay and integration between these functionalities. Some of the symbolic calculators studied in this book integrate a word-processor, spreadsheet, and dynamic geometry software, with means of interconnecting calculators, and calculator with computer. It is a reasonable hypothesis that these prefigure the scholastic tools of tomorrow. This makes it even more important to conceptualize didactic exploitations of those tools which support students in developing a construction of such devices as *coordinated systems of instruments*. In terms of researching and analyzing such usages, the work reported in this book is merely anticipatory. What we hope to have offered is a (tentative) system of ideas capable of taking into account both the constraints and affordances of future technological tools, and of helping to conceive appropriate pedagogical resources. These are necessary preconditions for integrating this wider and more efficient set of tools into new forms of teaching for effective learning.

This area of research, then, needs to be pursued further. It offers exciting prospects for informing and exploring new forms of integration of ICT by teachers. We see innovation and research as mutually supportive components of such development, calling for communication between practitioners and researchers. We hope that this book contributes to such work and encourages it further.

REFERENCES

Allen R., Cederberg J. & Wallace M. (1994) Teachers Empowering Teachers: A Professional Enhancement Model, *NCTM Yearbook: Professional Development for Teachers of Mathematics, National Council of Teachers of Mathematics*. Reston, VA.

Allen R., Wallace M., Cederberg J. & Pearson D. (1996) Teachers Empowering Teachers: Vertically-Integrated, Inquiry-Based Geometry in School Classrooms, *Mathematics Teacher* 90(3), 254-255.

Chevallard Y. (1992) Intégration et viabilité des objets informatiques, in B. Cornu (Ed.), *L'ordinateur pour enseigner les mathématiques* (pp. 183-203). Paris: PUF.

Guin D., Delgoulet J. & Salles J. (2000) Formation aux TICE: concevoir un dispositif d'enseignement autour d'un fichier rétroprojetable, *L'enseignement des mathématiques dans les pays francophones, Actes du colloque international EM2000*, CD-ROM. Grenoble: IREM, Université de Grenoble.

Guin D., Joab M. & Trouche L. (2003) *Suivi de formation à distance pour les enseignants: bilan de la phase expérimentale (2000-2002)*, CD-ROM. Montpellier: IREM, Université Montpellier II.

Kahane J. (2002) *L'enseignement des sciences mathématiques*, Centre National de Documentation Pédagogique. Paris: Odile Jacob.

Laborde C. (1999) Vers un usage banalisé de Cabri-Géomètre en classe de seconde: analyse des facteurs de l'intégration, *Calculatrices symboliques et géométriques dans l'enseignement des mathématiques, Actes du colloque francophone européen* (pp. 79-94). Montpellier: IREM, Université Montpellier II.

Laborde C. (2003) The design of curriculum with technology: lessons from projects based on dynamic geometry environments, *CAME Meeting*. Reims: IUFM de Reims.

Mariotti M. (2002) The influence of Technological Advances on Students' Mathematics Learning, in L. English, M.G. Bartolini Bussi, G. Jones, R. Lesh & D. Tirosh (Eds.), *Handbook of International Research in Mathematics Education* (pp. 695-723), Lawrence Erbaum Associates.

Ruthven K. & Chaplin D. (1997) The calculator as a cognitive tool: Upper-primary pupils tackling a realistic number problem, *International Journal of Computers for Mathematical Learning* 2(2), 93-124.

Ruthven K. (2001) Towards a new numeracy: the English experience of a 'calculator-aware' number curriculum, in J. Anghileri (Ed.), *Principles and Practice in Arithmetic Teaching* (pp. 165-188). Buckingham: Open University Press.

Vivet M. (1991) Usage des tuteurs intelligents: prise en compte du contexte, rôle du maître, in M. Baron, R. Gras & J.F. Nicaud (Eds.), *Actes des Deuxièmes journées EIAO* (pp. 239-246). Cachan: ENS.

Wenger E. (1998) *Communities of practice: Learning, meaning, and identity.* New York: Cambridge University Press.

INDEX

Mathematics Education Library

Managing Editor: A.J. Bishop, Melbourne, Australia

1. H. Freudenthal: *Didactical Phenomenology of Mathematical Structures.* 1983 ISBN 90-277-1535-1 HB; 90-277-2261-7 PB
2. B. Christiansen, A. G. Howson and M. Otte (eds.): Perspectives on Mathematics Education. Papers submitted by Members of the Bacomet Group. 1986 ISBN 90-277-1929-2; Pb 90-277-2118-1
3. A. Treffers: Three Dimensions. A Model of Goal and Theory Description in Mathematics Instruction TheWiskobas Project. 1987 ISBN 90-277-2165-3
4. S. Mellin-Olsen: The Politics of Mathematics Education. 1987 ISBN 90-277-2350-8
5. E. Fischbein: Intuition in Science and Mathematics. An Educational Approach. 1987 ISBN 90-277-2506-3
6. A.J. Bishop: Mathematical Enculturation. A Cultural Perspective on Mathematics Education. 1988 ISBN 90-277-2646-9; Pb (1991) 0-7923-1270-8
7. E. von Glasersfeld (ed.): Radical Constructivism in Mathematics Education. 1991 ISBN 0-7923-1257-0
8. L. Streefland: Fractions in Realistic Mathematics Education. A Paradigm of Developmental Research. 1991 ISBN 0-7923-1282-1
9. H. Freudenthal: Revisiting Mathematics Education. China Lectures. 1991 ISBN 0-7923-1299-6
10. A.J. Bishop, S. Mellin-Olsen and J. van Dormolen (eds.): Mathematical Knowledge: Its Growth Through Teaching. 1991 ISBN 0-7923-1344-5
11. D. Tall (ed.): Advanced Mathematical Thinking. 1991 ISBN 0-7923-1456-5
12. R. Kapadia and M. Borovcnik (eds.): Chance Encounters: Probability in Education. 1991 ISBN 0-7923-1474-3
13. R. Biehler, R.W. Scholz, R. Sträßer and B. Winkelmann (eds.): Didactics of Mathematics as a Scientific Discipline. 1994 ISBN 0-7923-2613-X
14. S. Lerman (ed.): Cultural Perspectives on the Mathematics Classroom. 1994 ISBN 0-7923-2931-7
15. O. Skovsmose: Towards a Philosophy of Critical Mathematics Education. 1994 ISBN 0-7923-2932-5
16. H. Mansfield, N.A. Pateman and N. Bednarz (eds.): Mathematics for Tomorrow's Young Children. International Perspectives on Curriculum. 1996 ISBN 0-7923-3998-3
17. R. Noss and C. Hoyles:Windows on Mathematical Meanings. Learning Cultures and Computers. 1996 ISBN 0-7923-4073-6 HB; 0-7923-4074-4 PB

Mathematics Education Library

18. N. Bednarz, C. Kieran and L. Lee (eds.): Approaches to Algebra. Perspectives for Research and Teaching.1996 ISBN 0-7923-4145-7; PB ISBN 0-7923-4168-6
19. G. Brousseau: Theory of Didactical Situations in Mathematics. Didactique des Mathématiques 19701990. Edited and translated by N. Balacheff, M. Cooper, R. Sutherland and V. Warfield. 1997 ISBN 0-7923-4526-6
20. T. Brown: Mathematics Education and Language. Interpreting Hermeneutics and Post-Structuralism. 1997 ISBN 0-7923-4554-1
    Second Revised Edition. 2001 Pb ISBN 0-7923-6969-6
21. D. Coben, J. O'Donoghue and G.E. FitzSimons (eds.): Perspectives on Adults Learning Mathematics. Research and Practice. 2000 ISBN 0-7923-6415-5
22. R. Sutherland, T. Rojano, A. Bell and R. Lins (eds.): Perspectives on School Algebra. 2000 ISBN 0-7923-6462-7
23. J.-L. Dorier (ed.): On the Teaching of Linear Algebra. 2000 ISBN 0-7923-6539-9
24. A. Bessot and J. Ridgway (eds.): Education for Mathematics in the Workplace. 2000 ISBN 0-7923-6663-8
25. D. Clarke (ed.): Perspectives on Practice and Meaning in Mathematics and Science Classrooms. 2001 ISBN 0-7923-6938-6; PB ISBN 0-7923-6939-4
26. J. Adler: Teaching Mathematics in Multilingual Classrooms. 2001 ISBN 0-7923-7079-1 HB; ISBN 0-7923-7080-5 PB
27. G. de Abreu, A.J. Bishop and N.C. Presmeg (eds.): Transitions Between Contexts of Mathematical Practices. 2001 ISBN 0-7923-7185-2
28. G.E. FitzSimons:What Counts as Mathematics? Technologies of Power in Adult and Vocational Education. 2002 ISBN 1-4020-0668-3
29. H.Alrø andO. Skovsmose:Dialogue and Learning in MathematicsEducation. Intention, Reflection, Critique. 2002 ISBN 1-4020-0998-4
30. K. Gravemeijer, R. Lehrer, B. van Oers and L. Verschaffel (eds.): Symbolizing, Modeling and Tool Use in Mathematics Education. 2002 ISBN 1-4020-1032-X
31. G.C. Leder, E. Pehkonen and G. Tőrner (eds.): Beliefs: A Hidden Variable in Mathematics Education? 2002 ISBN 1-4020-1057-5; PB ISBN 1-4020-1058-3
32. R. Vithal: In Search of a Pedagogy of Conflict and Dialogue for Mathematics Education. 2003 ISBN 1-4020-1504-6
33. H.W. Heymann: Why Teach Mathematics? A Focus on General Education. 2003 ISBN 1-4020-1786-3
34. L. Burton: Mathematicians as Enquirers: Learning about Learning Mathematics. 2004 ISBN 1-4020-7853-6 HB; 1-4020-7859-5 PB
35. P. Valero, R. Zevenbergen (eds.): Researching the Socio-Political Dimensions of Mathematics Education: Issues of Power in Theory and Methodology, 2004 ISBN 1-4020-7906-0
36. D. Guin, K. Ruthven, L. Trouche (eds.) The Didactical Challenge of Symbolic Calculators: Turing a Computational Device into a Mathematical Instrument 2005 ISBN 0-387-23158-7

Printed in the United Kingdom
by Lightning Source UK Ltd.
116871UKS00004B/179

9 780387 231587